Analysis of Two-Way Layouts

ANALYSIS OF TWO-WAY LAYOUTS

JOHN MANDEL

Guest Researcher National Institute of Science and Technology

CHAPMAN & HALL

I(T)P An International Thomson Publishing Company

New York • Albany • Bonn • Boston • Cincinnati
• Detroit • London • Madrid • Melbourne • Mexico City
• Pacific Grove • Paris • San Francisco • Singapore
• Tokyo • Toronto • Washington

Cover Design: Andrea Meyer, EmDash Inc.

Copyright © 1995
By Chapman & Hall
A division of International Thompson Publishing Inc.
I(T)P The ITP logo is a trademark under license

Printed in the United States of America

For more information, contact:

Chapman & Hall
One Penn Plaza
New York, NY 10119

Chapman & Hall
2-6 Boundary Row
London SE1 8HN

International Thompson Publishing Europe
Berkshire House 168-173
High Holborn
London WC1V 7AA
England

International Thomson Editores
Campos Eliseos 385, Piso 7
Col. Polanco
11560 Mexico D.F. Mexico

International Thomson Publishing Gmbh
Königwinterer Strasse 418
53227 Born
Germany

Thomas Nelson Australia
102 Dodds Street
South Merlbourne, 3205
Victoria, Australia

International Thomson Publishing Asia
221 Henderson Road
#05-10 Henderson Building
Singapore 0315

Nelson Canada
1120 Birchmount Road
Scarborough, Ontario
Canada, M1K 5G4

International Thomson Publishing Japan
Hirakawacho-cho Kyowa Building, 3F
2-2-1 Hirakawacho-cho
Chiyoda-ku, Tokyo 102
Japan

1 2 3 4 5 6 7 8 9 10 XXX 01 00 99 97 96 95

Library of Congress Cataloging-in-Publication Data

Mandel, John, 1914-
    Analysis of two-way layouts / John Mandel.
       p.   cm.
    Includes bibliographical references and index.
    ISBN 0-412-98611-6
    1. Statistics—Charts, diagrams, ect. 2. Tables (Systematic lists) I. Title.
QA276.25.M34 1994

001.4'225—dc20

94-48326
CIP

Please send your order for this or any Chapman & Hall book to **Chapman & Hall, 29 West 35th Street, New York, NY 10001, Attn: Customer Service Department.** You may also call our Order Department at 1-212-244-3336 or fax your purchase order to 1-800-248-4724.

For a complete listing of Chapman & Hall's titles, send your requests to **Chapman & Hall, Dept. BC, One Penn Plaza, New York, NY 10119.**

Contents

Contents

14. Applications to Real Data Sets

15. One-Way Tables and the *h* Statistic

16. How Useful Are Models?

Preface

This book deals with data that appear as two-way tables, that is, rectangular arrays of rows and columns. The intersection of a row and a column is called a cell. In each cell there appears either a single measurement or several "replicate" measurements. The rows and columns of the table may correspond to quantitative entities such as volume, pressure, temperature, year, and month, or to qualitative entities such as laboratories, operators, samples, patients, and so on.

We use the word "replicate" in the usual sense: as repeated measurements made for the same values in the row and in the column. This is also the way that chemists and most scientific experimenters understand the term. For example, if the rows represent *temperature*, the columns *pressure*, and the entry of the table is *volume*, then replicates are measurements of volume made at a specific temperature (row) and a specific pressure (column). This is not the way the term "replicate" is used in certain statistical texts, namely, as repetitions of the entire experiment (all rows and all columns).

We are interested in the *analysis* of the two-way data. It is important to explain what is meant by "analysis." Contrary to a widespread belief, the fitting of a *model* to a data set does *not*, by itself, constitute an *analysis* of the set, but often it is not even a necessary part of it. We will return to this matter in Chapter 16. Suffice it to say now that analysis in our sense is an elucidation of all that we can learn from the data, regardless of how we do it, as long as our method is rationally defensible.

Data are the result of measurements made under more or less controlled conditions in what is usually called an *experiment*. Experiments are carried out in order to learn something about a certain subject. The crucial point is that whatever we learn from the experiment is clearly stated, as a result of the analysis of the data, in terms understandable to the experimenter. Often the data are puzzling. Then we should investigate what bothers us in them. We can, of course, manipulate data, but we should always remember that data are, as it were, *messengers* who tell us something about the physical experiment. Therefore, actions taken with respect to the data (for example, the rejection of "outliers") are meaningless and, indeed, harmful unless these actions are also understood and implemented in terms of what happened in the experiment. This brings us to the "modeling" of the data. In this book, many models will be discussed, although we prefer to use the term *structure*. We do this primarily to show that the usual ways of analyzing two-way data (mostly by analysis of variance) are very often wrong because they *impose* a wrong structure on the data. Statisticians have been aware of this to a certain extent and have proposed transformations of scale of various sorts to *bring the data into harmony with the model*. We believe that the opposite should be done: *Bring the model into harmony with the data*; in other words, use a model that is adapted to the data regardless of whether it can be expressed as one of the usual categories of statistical textbooks.

We do not believe that the structures we discuss are necessarily the "true" structure of the data: they are often only a reasonable approximation. It follows that if we draw inferences from the fitted structure rather than from the original data, we may well miss important aspects of the data and of the experiment. We will illustrate this in Chapter 16 in terms of a real data set. If you make a graph with fitted data only, then make sure to compare it to the same graph made on the original data and observe whether a difference exists between the two graphs.

The main objective of this book is to present a number of structures that are useful in the analysis of two-way tables, emphasizing at the same time the limitations of any model for a thorough analysis of the data sets. It will also be seen that data can often be analyzed without recourse to any preconceived model, and how this can be done.

A question that may arise in the mind of the reader is: Why two-way tables? What is so special about two-way tables to warrant the writing of a book about them? The answer is philosophically important. A one-way table is just a string of numbers. The only information contained in this string, apart from the magnitude of each number, is the location which it occupies in the string. Any "structure" that the string possesses must be found in the relation between the location and the number. If you destroy this relation, for example, by randomizing the set, the structure is gone.

Not so for two-way tables. A rectangular array of numbers has, of course, rows and columns. These may carry a "label", for example, the rows representing temperature and the columns pressure. The analyst is interested in the relationship between the numbers *inside* the table and these labels. But, *aside from these relationships*, the numbers inside the table may exhibit structure. You can erase the labels in the rows and the columns, and then randomize the rows and randomize the columns. All this will *not* destroy this structure. We will call this the *internal* structure of the table. It will become apparent to the reader that the internal structure of a two-way table is a matter of fundamental interest. Thus, a basic *qualitative* difference exists between one-way and two-way tables.

We have limited our discussions to two-way tables. What is true for two-way tables, in terms of internal structure, is also true for tables of higher dimensionality, such as three-way, four-way tables, and so on. But we believe that the complexity of two-way tables in terms of structure is already very great. The study of structure in tables of higher dimensionality is certainly justified, but it is a subject beyond the scope of this book.

We hope that the book will induce scientists to deal with data in a thorough and logical way, rather than to apply textbook methods of analysis.

Introduction

To assist the reader, we present in the following a quick review of the contents of its chapters.

Chapter 1 is a general review of three cases for the fitting of straight lines to data. The reason for this preliminary chapter is that much of the book requires the fitting of straight lines. The three cases are the "classical" one, the "weighted" case, and the so-called "errors in variables" case, that is, the case in which both x and y are affected by experimental errors.

In Chapter 2, we introduce the concept of structure of a two-way table. We discuss the additive, the multiplicative, the row-linear, and the column-linear structures.

Chapter 3 introduces general functional expressions for the structures discussed in the previous chapter. It also examines the effect that coding of data has on their structure.

In Chapter 4, we present a detailed examination of the concurrent structure and a necessary and sufficient condition for concurrence. We also discuss the interdependence of cell y_{ij} and the corresponding column average x_i, and the effect of this interdependence on an analysis of two-way tables.

Chapter 5 contains a discussion of h and h' statistics, and their plots. These are useful tools for the examination of two-way tables.

Chapter 6 shows how the sum of squares of residuals from various fits can be interpreted by means of a theorem on structures embedded inside other structures.

Chapter 7 introduces a very general model for two-way tables, consisting of an additive part and a nonadditive part. The latter consists of a sum of products, each product being that of a row-dependent quantity by a column-dependent quantity. The chapter also presents an algorithm for the calculation of the product terms.

Chapter 8 applies the procedures presented in previous chapters to the analysis of several sets of data.

Chapter 9 deals with tables in which each cell contains several replicate measurements. It introduces the k statistic for the evaluation of within-cell variability.

Chapter 10 deals with interlaboratory testing. This comprises proficiency studies, for the purpose of testing the laboratories, and studies of the repeatability and reproducibility of test methods.

Chapter 11 explains what can be done with two-way tables in which one or more cells are missing.

Chapter 12 deals with curve fitting. It introduces the Quadratic Four-Parameter Curve (QFP), which is a generalization of the quadratic. A method for fitting QFP to a set of data is explained in detail.

In Chapter 13, the two-stage procedure for fitting a two-way table is explained in terms of an example taken from the literature.

Chapter 14 presents six examples taken from the literature. A complete analysis is presented for five of the example. For one of the six, the analysis is incomplete, for reasons explained in this chapter.

Chapter 15, unlike the bulk of the book, deals with one-way tables and illustrates the use of the h statistic for their analysis.

Chapter 16 discusses the advantages and disadvantages of "modeling" data. It shows, by means of an example, that a model is not always necessary for a meaningful analysis of the data.

1

Statistics of the Straight Line

1.1 The Classical Case

In this book the straight line will play a major role. Therefore, we begin by recalling the essential results of the theory of fitting straight lines, often referred to as linear regression. We first discuss the classical case.

Consider a set of x- and y-values, as shown in Table 1.1. Let x_i and y_i represent the ith pair, out of p pairs. In Table 1.1, $p = 7$. In a plot of the data, x_i will be the abscissa and y_i the ordinate of the ith point. The plot of Table 1.1 is shown in Fig. 1.1.

Table 1.1. Data for a Straight Line Fit[a], Classical Case

x_i	y_i
10	0.959652
15	0.956724
20	0.953692
25	0.950528
30	0.947259
35	0.943874
40	0.940390

SOURCE: Wood and Martin (1964)

[a]Density of aqueous solutions of ethyl alcohol (30% alcohol) at various temperatures (x = temperature in degrees Celsius; y = density of alcohol).

The equation we consider is

$$y_i = a + \beta x_i + \varepsilon_i .$$
(1.1)

Except for the presence of the experimental error ε_i, Eq. (1.1) represents a straight line with the y-intercept equal to a and slope equal to β. We assume in the classical case that x is not subject to experimental error, but that y consist of two parts: the expected value of y, denoted by $E(y)$, and an error ε_i:

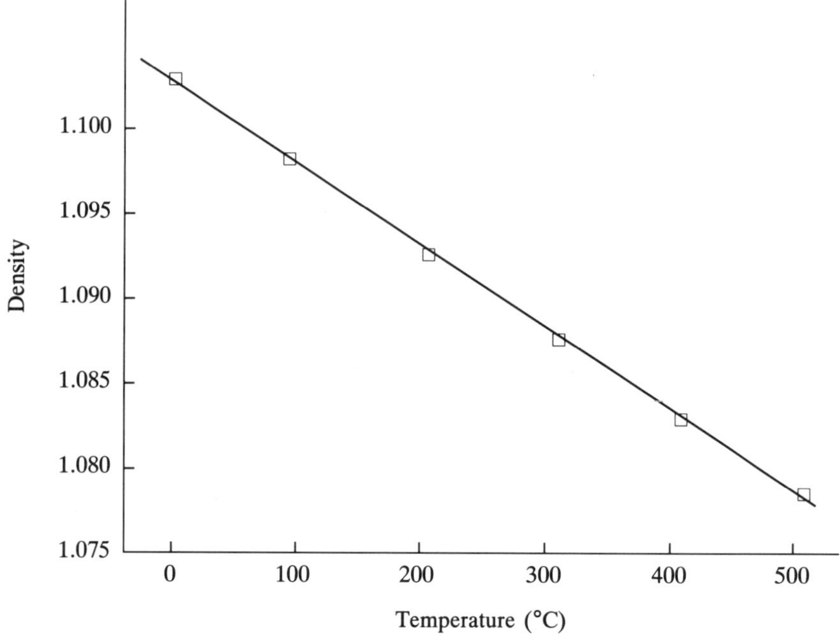

Figure 1.1. Density of aqueous solutions of ethyl alcohol.

$$y_i = E(y_i) + \varepsilon_i . \tag{1.2}$$

We also assume that the experimental errors ε_i are independent of each other and that they have a common variance, which we denote by $\mathrm{Var}(\varepsilon)$ or by $(\sigma_\varepsilon)^2$.

1.2 Computations for the Classical Case

Under these conditions the following applies.

We first define the *averages* of x_i and y_i:

$$\bar{x} = \Sigma_i(x_i)/p \qquad \text{and} \qquad \bar{y} = \Sigma_i(y_i)/p \tag{1.3}$$

Then we define the quantities *XX*, *YY*, and *XY* as follows:

$$XX = \Sigma_i(x_i - \bar{x})^2 , \tag{1.4a}$$

$$YY = \Sigma_i(y_i - \bar{y})^2 , \tag{1.4b}$$

$$XY = \Sigma_i(x_i - \bar{x})(y_i - \bar{y}) . \tag{1.4c}$$

An estimate b of the slope β is given by

$$b = XY/XX. \tag{1.5}$$

An estimate a of the intercept α is given by

$$a = \bar{y} - b\bar{x}. \tag{1.6}$$

Because of the error ε_i, the points on the plot in Fig. 1.1 do not lie *exactly* on the line. The scatter of the points about the line is measured by $(\sigma_\varepsilon)^2$, which is not known.

We can, however, obtain an estimate of $(\sigma_\varepsilon)^2$, which we denote by $(s_\varepsilon)^2$, as follows: Let

$$\hat{y}_i = a + bx_i. \tag{1.7}$$

The difference $y_i - \hat{y}_i$ is called a *residual*:

$$d_i = y_i - \hat{y}_i. \tag{1.8}$$

This is the vertical distance between the *observed* point (x_i, y_i) and the corresponding *fitted* point $(x_i, \hat{y}_i)$. The estimate $(s_\varepsilon)^2$ is given by

$$(s_\varepsilon)^2 = \frac{\Sigma_i (d_i)^2}{p - 2}. \tag{1.9}$$

The reason for the denominator $p - 2$ is that two degrees of freedom are "used up" in the estimation of α and β

The variance of the estimate b is given by

$$(s_b)^2 = (s_\varepsilon)^2 / XX \tag{1.10}$$

and the variance of a is given by

$$(s_a)^2 = (s_\varepsilon)^2 \left(\frac{1}{p} + \frac{\bar{x}^2}{XX} \right). \tag{1.11}$$

The a and b estimators (general formulas for calculating a and b) are correlated with each other. Therefore, we prefer to write the equation for the fitted value of y as

$$\hat{y}_i = \bar{y} + b(x_i - \bar{x}), \tag{1.12}$$

where $\bar{x}$ and $\bar{y}$ are the quantities defined in Eq. (1.3). The estimators $\bar{y}$ and b are statistically independent (of each other) provided that the ε_i are independent and normally distributed. We refer to $\bar{y}$ as the "height" of the fitted line, and note that the point $(\bar{x}, \bar{y})$ lies on the fitted line. Clearly, $\bar{y}$ is the ordinate on the fitted line corresponding to the abscissa $\bar{x}$.

1.3 An Example of Classical Linear Regression

In Table 1.2, the fitted y, $\hat{y}$, and the residual d are displayed for each x-value. $\bar{x}$, $\bar{y}$, XX, YY, XY, the intercept a, the slope b, and the standard deviation of the residuals, s_ε, are also listed.

Table 1.2. Fit of Straight Line to Table 1.1 Data

x	$\hat{y}$	d	
10	0.959937	-0.000285	$\bar{x} = 25$
15	0.956726	-0.000002	$\bar{y} = 0.950303$
20	0.953514	0.000178	$XX = 700$
25	0.950303	0.000225	$YY = 0.000289034$
30	0.947092	0.000167	$XY = -0.449595$
35	0.943880	-0.000006	
40	0.940669	-0.000279	

$a = 0.96636,\quad b = -0.000642279,\quad \bar{y} = 0.950303,\quad s_\varepsilon = 0.000232$

The presence of residuals in this table is of crucial importance. With the advent of computers, or even good hand-held calculators, the calculation of the residuals is extremely simple and rapid. To omit them is to deprive oneself of crucial information about the adequacy of the fit. Note also that the x-values have been ordered from low to high. This again is a simple, but important step for better understanding the fit. Figure 1.1 appears to inform us that the fit is very good. However, an examination of the residuals in Table 1.2 reveals curvature. Indeed, the succession of signs of the residuals $(--+++--)$ indicates points that start *below* the fitted line, rise *above* it, and then fall *below* it.

Thus, the appearance of the graph of the fitted line, together with the observed points, is not always a good indication of the adequacy of the fit, especially for data of high precision such as those of Table 1.1. Therefore, a study of residuals is essential.

1.4 Weighted Straight Line Fitting

The procedure discussed above is based on the assumption that the standard error of each y_i is the same. This assumption is, of course, not always satisfied. There are cases in which the standard errors of the p points, although different from each other, are nevertheless known, at least relatively to each other.

Consider the data in Table 1.3. They are taken from Fisher (1941) and represent the number of eye facets of Drosophila melanogasta, in factorial

Table 1.3. Eye Facets of Drosophila Melanogasta

Temperature	Average Count	Number of Items
15	4.5478	90
17	3.5144	54
19	2.4917	83
21	0.3200	100
23	-2.4300	86
25	-1.6267	122
27	-4.6234	137
29	-6.6596	98
31	-7.7130	53

units, for a set of nine temperatures from 15°C to 31°C. The number of units varies from temperature to temperature, as shown in Table 1.3. Assuming a constant variance for each unit, the variance of each y_i is given by

$$(\sigma_{y_i})^2 = \sigma^2/n_i \,, \tag{1.13}$$

where n_i is the number of units at temperature x_i given in Table 1.3. A plot of the average count at each temperature versus temperature is shown in Fig. 1.2.

We define the (absolute) weight, w'_i, corresponding to point i as the reciprocal of the variance:

$$w'_i = n_i/\sigma^2 \,. \tag{1.14}$$

Because σ^2 is a constant, a set of *relative* weights is given by n_i, a set of numbers proportional to the absolute weights. It can be shown that relative weights are sufficient to estimate the parameters of a straight line of y versus x. The formulas are as follows. First, denote the relative weights by w_i. In the case of our example, $w_i = n_i = \sigma^2 w'_i$. In general,

$$w_i = K^2 w'_i \,,$$

$$\tag{1.15}$$

where K^2 is not necessarily known.

The *weighted* averages, denoted by x and y, are defined by

$$\bar{x} = \frac{\Sigma(w_i x_i)}{\Sigma w_i} \quad \text{and} \quad \bar{y} = \frac{\Sigma(w_i y_i)}{\Sigma w_i}. \tag{1.16}$$

The quantities XX, YY, and XY are now defined as

$$XX = \Sigma_i w_i (x_i - \bar{x})^2 \tag{1.17a}$$

$$YY = \Sigma_i w_i (y_i - \bar{y})^2 \tag{1.17b}$$

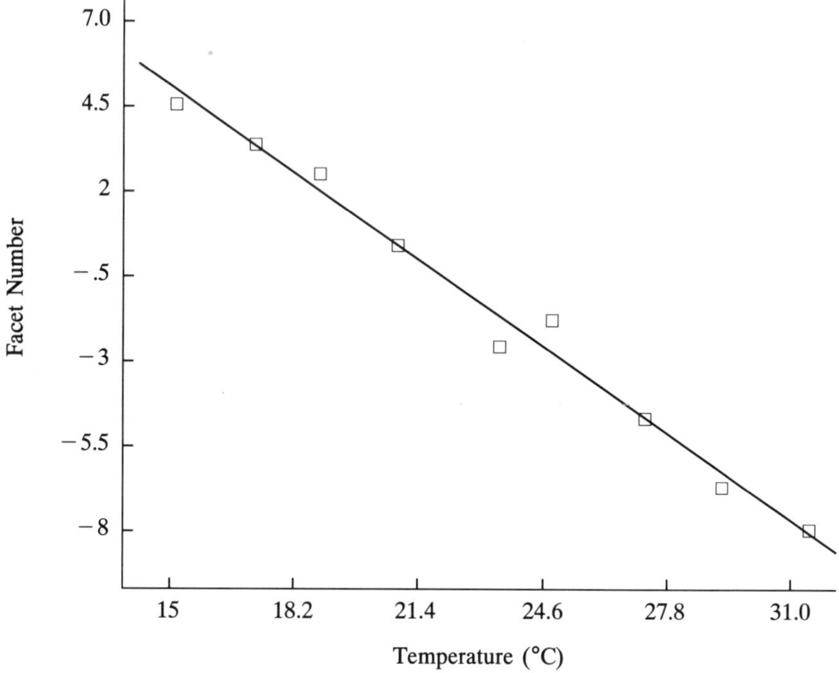

Figure 1.2. Eye facets versus temperature (Drosophila melanogasta).

$$XY = \Sigma_i \, w_i (x_i - \tilde{x})(y_i - \tilde{y}). \tag{1.17c}$$

The height is $\tilde{y}$; an estimate of the slope β, denoted as before by b, is

$$b = XY/XX. \tag{1.18}$$

The equation for the fitted line is

$$\hat{y} = \tilde{y} + b(x_i - \tilde{x}). \tag{1.19}$$

The ith residual, d_i, is

$$d_i = y_i - \hat{y}_i. \tag{1.20}$$

The estimated variance of a *single* unit is

$$s^2 = \frac{\Sigma_i (w_i \, d_i^2)}{p - 2}. \tag{1.21}$$

Thus, s^2 is an estimate of σ^2. According to Eqs. (1.14) and (1.15), $\sigma^2 = n_i/w_i' = n_i(w_i/K^2)^{-1} = K^2(n_i/w_i)$. However, because we used n_i as a relative weight, we have

$$w_i = n_i$$

and, consequently,

$$\sigma^2 = K^2 \,.$$

Therefore, s^2 is an estimate of K^2, and for the estimate $\widehat{K}^2$ of K^2, we write

$$\widehat{K}^2 = s^2 \,. \tag{1.22}$$

1.5 An Example of Weighted Linear Regression

Applying these formulas to our example, we have the results shown in Table 1.4.

Table 1.4. Weighted Straight Line Fit to Data of Table 1.3

x	$\hat{y}$	d	w	
15	5.0398	-0.4920	90	$\bar{x} = 23.2770$
17	3.4508	0.0636	54	$\bar{y} = -1.536309$
19	1.8618	0.6299	83	$XX = 18968.8338$
21	0.2728	0.0472	100	$YY = 12369.9044$
23	-1.3162	-1.1138	86	$XY = -15070.7424$
25	-2.9052	1.2785	122	
27	-4.4942	-0.1292	137	$s^2 = \dfrac{396.1964}{9-2} = 56.5995$
29	-6.0832	-0.5764	98	
31	-7.6722	-0.0408	53	
	$a = 16.9573,$	$b = -0.7945,$	$s = 7.5233$	

It is worth noting that whereas in the classical case $\Sigma_i \, d_i = 0$, in the weighted case $\Sigma_i \, (w_i \, d_i) = 0$. In the classical case, the fit is such that $\Sigma_i \, d_i^2$ is a minimum (among all linear unbiased fits); in the weighted case, it is $\Sigma_i \, (w_i \, d_i^2)$ that is minimum.

The value of s is 7.52, whereas the "internal" estimate of the standard deviation of a single unit, d [see Fisher (1941)], is 2.20. The F statistic, with 7 and 814 degrees of freedom, yields

$$F = \frac{(7.52)^2}{(2.20)^2} = 11.68 \,.$$

This value is highly significant. Thus, the weighting has *not* reduced the goodnes-of-fit to what we would expect from the internal measure of variability. The conclusion is that there exists an additional source of variability that affects the average counts, even after the linear relationship has been taken into account. Figure 1.2 suggest that the true relation is not linear, but no further conclusion can be drawn from the data without, possibly, the assistance of a subject-matter specialist.

1.6 Errors in Both Variables

In addition to the classical case and that of weighted regression, we consider a third possibility.

Suppose that both x and y are subject to measurement error, as shown by the following equations:

$$x_i = E(x_i) + \delta_i \qquad \text{and} \qquad y_i = E(y_i) + \varepsilon_i, \tag{1.23}$$

where $E(x_i)$ and $E(y_i)$ are the expected values (the true values) of x_i and y_i, respectively, δ_i and ε_i are the random errors. We assume that the variance of δ_i is $(\sigma_\delta)^2$ and the variance of ε_i is $(\sigma_\varepsilon)^2$. The variances are not known, but we assume that we *do* know their ratio, which we call λ. Thus,

$$\lambda = (\sigma_\varepsilon)^2/(\sigma_\delta)^2 \tag{1.24}$$

and λ is known.

Furthermore, we assume that a linear relation exists between $E(x_i)$ and $E(y_i)$:

$$E(y_i) = \alpha + \beta E(x_i). \tag{1.25}$$

Under these conditions, the problem can be solved [see, for example, Mandel (1984)]. The formulas are as follows:

The estimate b of β is given by

$$b = \frac{YY - \lambda XX + \sqrt{(YY - \lambda XX)^2 + (4\lambda XY^2)}}{2XY}, \tag{1.26}$$

where XX, YY, and XY are defined as in Eq. (1.4).

The estimate a of α is given by

$$a = \bar{y} - b\bar{x}. \tag{1.27}$$

We define a quantity d by

$$d_i = y_i - a - bx_i, \tag{1.28}$$

and a quantity D^2 by

$$D^2 = \frac{\Sigma d_i}{(p - 2)}, \tag{1.29}$$

where p is the number of (x, y) points. In addition, let us define

$$P = \lambda + b^2, \tag{1.30}$$

$$Q = \lambda^2 XX + 2\lambda b XY + b^2 YY. \tag{1.31}$$

Then we have for estimates of the standard deviations of δ and ε,

$$s_\delta = \frac{D}{\sqrt{P}} \quad \text{and} \quad s_\varepsilon = \frac{D\sqrt{\lambda}}{\sqrt{P}}, \tag{1.32}$$

and for the estimates of the standard deviation of a and b,

$$s_a = D\sqrt{\frac{1}{P} + \frac{\bar{x}^2 P^2}{Q}}, \tag{1.33}$$

$$s_b = \frac{DP}{\sqrt{Q}}. \tag{1.34}$$

Finally, estimates of the true values of x_i and y_i are given by

$$\hat{x}_i = x_i + \frac{b}{P} d_i \tag{1.35}$$

$$\hat{y}_i = y_i - \frac{\lambda}{P} d_i. \tag{1.36}$$

1.7 An Example of Regression with Errors in Both Variables

Table 1.5 presents data of the specific volume of rubber (Wood and Martin, 1964). We compare the specific volume of unvulcanized rubber (y-value) with that of cured rubber (x-value). Both measurements were made at $25°$ Celsius on six samples of rubber.

Table 1.5. Specific Volume of Rubber

x	y
1.0786	1.0702
1.0830	1.0748
1.0877	1.0795
1.0926	1.0845
1.0977	1.0896
1.1032	1.0951

Table 1.6. Specific Volume of Rubber. Fit of Straight Line to Data of Table 1.5

$XX = 0.000423233$	$YY = 0.000432468$	$XY = 0.000427607$
$\lambda = 1$	$a = 0.020023$	$b = 1.010857$
$P = 2.021832$	$Q = 0.001730$	

$$\bar{x} = 1.090467, \quad \bar{y} = 1.082283, \quad D = 0.000068$$

$s_\delta = 0.000048$	$s_a = 0.003620$
$s_\varepsilon = 0.000048$	$s_b = 0.003320$

$\hat{x}$	$\hat{y}$
1.07856	1.07024
1.08303	1.07477
1.08771	1.07949
1.09263	1.08447
1.09770	1.08960
1.10317	1.09513

In this case, the assignment of the symbols x and y to the two variables is arbitrary: All formulas are symmetric and an interchange of x and y would not affect the final results, provided an interchange of x and y is accompanied by a change of λ to $1/\lambda$. For our example, the ratios of the variances, λ, can be assumed to be unity, as the measurements were made in the same manner for both variables:

$$\lambda = 1.$$

Then the application of Eqs. (1.23)–(1.36) leads to the results shown in Table 1.6. A plot of the data is shown in Fig. 1.3.

1.8 The Importance of Choosing the Correct Model

It is generally believed that models in linear regression are chosen to obtain a better fit. Actually, the difference in the fit is in many cases negligible.

Why then select a model? The real reason is to obtain a better representation of the errors in one or both variables. The weighted expression allows us to arrive at a correct representation of the error structure of y, for the case in which the variance of y is not constant. The case of errors in both variables, likewise, is important because it allows us to allocate properly the errors in both x and y. As observed earlier, in this model it does not matter which variable is treated as x and which as y. This symmetry is often a very desirable situation.

Thus, for our example of the specific volume of rubber, there is no reason

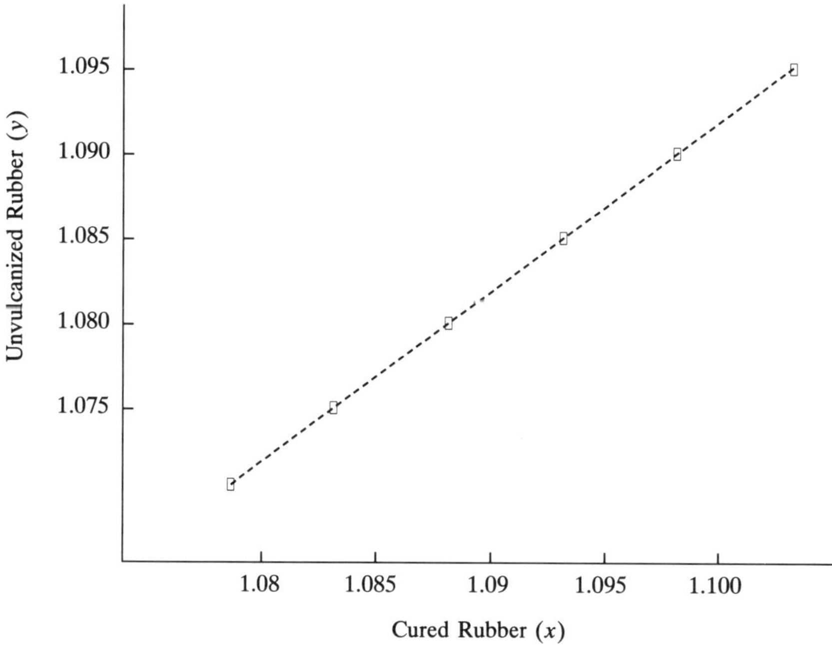

Figure 1.3. Specific volume of rubber, unvulcanized versus cured. Data of Table 1.5.

to call either variable the independent variable and the other the dependent variable. Of course, if x and y are interchanged, then as mentioned earlier, λ becomes $1/\lambda$.

References

Fisher, R.A. (1941). *Statistical Methods for Research Workers*. G.E. Stechert & Co., New York, Chap. VIII.

Mandel, John (1984). Fitting straight lines when both variables are subject to error. *J. Qual. Technol.* 16, 1–14.

Wood, L.A. and G.M. Martin (1964). Compressibility of natural rubber at pressures below 500 kg/cm^2. *J. Res. Nat. Bur. Stand., Sec. A. Phys. Chem.*, 68A, 259–268.

2
Structures of Two-Way Tables

2.1 Introduction and Terminology

There are numerous instances, in the physical sciences as well as in other disciplines, in which experimental data appear in the form of two-way tables. Let us just mention a few examples. In chemistry, equations of state are relations between volume, pressure, and temperature. Thus, volume can be tabulated in a two-way table in which the rows represent pressure (or the related quantity density) and the columns represent temperature.

In physics, the density of an aqueous solution of alcohol is a function of temperature and the concentration of alcohol in the solution.

In animal husbandry, the percent protein in the milk of cows is a function of the diet administered to the cows and of the duration during which the diet is given.

In all of these cases, it is important to study the relationship in question, that is, the relation between the two quantities representing the rows and the columns of the table, considered as "independent" variables, and the tabulated quantity which is considered the "dependent" variable.

We will use a different terminology, because "independent" and "dependent" can occasionally be misleading. In accordance with common statistical methodology, the "independent" variables are generally called "regressors" and the "dependent" variable is called "response." If a relationship is established between the regressors on the one hand and the response on the other hand, this relationship is said, in statistics, to be a "regression" of the response on the regressors.

In this book, we will use still another terminology, which is more directly related to two-way tables. The quantities representing rows and columns will be referred to as "row labels" and "column labels." Thus, the regression becomes one between the labels and the response.

The reason for calling the independent variables "labels" is twofold. In many cases one, or both, of the labels may not be measurable quantities at all; they may be days, laboratories, instruments, methods, or more generally categories.

Secondy, we will present a two-step method of analysis. The first step is to relate the response, *not* to the labels, but to the row and column averages of the response; the second step (if applicable) relates these row and column *averages* to the row and column *labels*, respectively.

Let us clarify this by means of an example. In Table 2.1, the quantity "specific volume" is tabulated in a two-way table (Wood and Martin, 1964). The rows of the table represent different "levels" (values) of the quantity "temperature," which is the row label, and the columns of the table represent different levels of the quantity "pressure," which is the column label. Moreover, we have calculated and listed the average of the values in each row (the row averages) and the average of the values in each column (the column averages).

Table 2.1. Specific Volume of Peroxide-Cured Rubber [a]

Temp. (°C)	Pressure (kg/cm²)						Avg.
	500	400	300	200	100	0	
0	137	178	219	263	307	357	243.50
10	197	239	282	328	376	427	308.17
20	256	301	346	394	444	498	373.17
25	286	330	377	426	477	532	404.67
Avg.	219.00	262.00	306.00	352.75	401. 00	453.50	332.38

SOURCE: Wood and Martin (1964).
[a] Tabulated response = (specific volume $-1.05) \cdot 10^4$.

It is seen that we can relate each row average to *its* corresponding row label (for example, 308.17 to a temperature of 10°) and each column average to *its* corresponding column label (for example, 352.75 to a pressure of 200 kg/cm²).

Our two-step procedure consists in first relating the response to the row and column averages, and then, relating the row averages to row labels, and the column averages to column labels. If the row labels are not a measured quantity, but rather a *category* (such as laboratories), then the second step is not possible for the rows. A similar situation may hold for the columns.

We will often refer to the labels as marginal labels because they appear in the margins of the table.

2.2 Structure

In this section we will deal with two-way tables, but we will temporarily ignore the labels altogether. This leaves us with the response and the row and column averages.

Let us decide here, and for the rest of this book, that the number of rows is represented by the letter p and the number of columns is represented by the number q. Thus, the table consist of pq (p times q) values of the response. To this we may add p row averages and q column averages, but these are values *calculated* from the pq response values.

We consider Table 2.2 which contains four tables of responses. In Table 2.2a, we observe that the *difference* between two response-values that occur in the same column but in different rows, say the first and third rows, is 7, *regardless of which column was picked.* A similar situation holds if we consider the *difference* between the response values in the same row but in different columns. It is easily proved that if this condition holds for rows, then it holds for columns, and vice versa.

Table 2.2. Four Sets of Artificial Data

(a)				(b)			
10	12	9	11	3	2	4	1
6	8	5	7	6	4	8	2
3	5	2	4	9	6	12	3
(c)				(d)			
3	2	4	1	9.77	10.22	8.86	10.40
7	5	9	3	6.18	8.02	4.28	7.13
1	4	-2	7	3.01	4.85	2.17	3.88

This condition of *constant difference* is called *additivity* and we may say that we are dealing with an *additive* two-way table. We can also express it by saying that the *model* underlying the values is *additive*.

Clearly, additivity is a very special condition, and it would be surprising if we encountered many additive tables in real life. However, a set of data may still be additive, even if the condition of constant differences does not strictly hold. This comes about as a result of *experimental error*.

Suppose that to each response value in Table 2.2a we *add* an uncertainty value ε to simulate the experimental error which is always present in actual measurements. The values are shown in Table 2.2d. The pq values for ε are supposed to be *random* drawings from a *population* of such numbers. Evidently, after adding the ε values, the additivity feature will no longer hold. Nevertheless, we still refer to the set as additive if certain conditions hold. This will be discussed in detail later.

For the moment, we consider response values *not* affected by experimental error.

Turning to set 2.2b (Table 2.2b) we easily verify that it is *not* additive. However, if in set 2.2b we consider not differences, but *quotients* of numbers, we find again a striking feature. Thus, we find that the *quotient* of two response values occurring in the same column but in different rows is the same, regardless of which columns we pick. Again, the same condition holds when we interchange rows and columns. Here too, if we add experimental errors to the values, they would, to a certain extent, obscure the constant quotient relationship. For reasons to be explained later we call set 2.2b a *multiplicative* set.

Set 2.2c is neither additive nor multiplicative. Nevertheless, even this set obeys a definite structure, and we will learn how to detect and characterize this structure.

It is important to reiterate that the structure types we have discussed are *internal*. By that we mean that they exist without any reference to the labels of the table. We will refer to any relation involving the labels as *external*. We also recall that the structures we have encountered exist *exactly* only because of the absence of experimental error.

2.3 The Row-Linear Structure

We consider Table 2.3, which consist of artificial data. At the bottom of the table we have added the column-averages; which we have denoted by x_j. There are no labels in this table.

Table 2.3. An Artificial Set of Data

	16.16	23.60	9.39	27.67	12.13
	9.87	15.01	4.88	18.31	7.25
	12.04	6.06	18.04	2.67	15.58
	9.60	14.39	4.79	17.19	6.40
Avg.	11.918	14.765	9.275	16.460	10.340

In Figure 2.1, we have plotted each of the four rows of the table against the x_j values. It is apparent that the points corresponding to each row are *close* to, but not exactly on, a straight line; it is not unreasonable to assume that they would lie *exactly* on straight lines if they were not affected by experimental errors. We also see that the four straight lines are different from each other. A straight line is characterized by two parameters: its y-intercept and its slope. We see that our lines differ from each other *both* in intercept and slope. The reader can easily verify that this condition of straight line relations is by no means true for all two-way tables. It is, rather, indicative of a special structure which we will refer to as a *row-linear* structure. Let us write an equation for this structure. If y_{ij} represents the test result in row i and column j, then we have

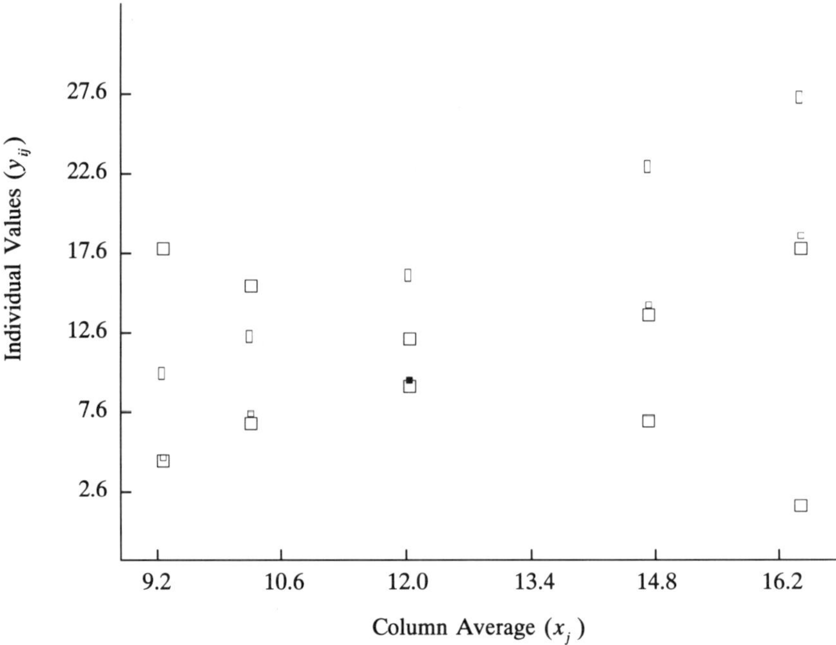

Figure 2.1. A row-linear set (artificial); y_{ij} versus x_j. Data of Table 2.3.

$$y_{ij} = a_i + \beta_i x_j + \varepsilon_{ij}. \tag{2.1}$$

For any given i (row), this equation expresses a linear relation between x_j and y_{ij}, marred only by the error ε_{ij}.

If ε_{ij} was not present in this equation, then the relation would be *exactly* linear. In that case, all of the points for each row would lie exactly on a straight line. The ε_{ij} cause the points to scatter about this line; the larger the ε_{ij} the more scatter will appear.

Note that a_i and β_i apply to row i. They may differ from row to row. In accordance with the theory of linear regression (Chap. I), the estimate b_i of β_i is obtained by

$$b_i = \frac{\Sigma_j (y_{ij} - \bar{y}_i)(x_j - \bar{x})}{\Sigma_i (x_j - \bar{x})^2}, \tag{2.2}$$

where the bar (e.g. $\bar{x}$, $\bar{y}_i$) indicates an average (in this case, over j).

An estimate of a_i, denoted by a_i, is obtained by

$$a_i = \bar{y}_i - b_i \bar{x}. \tag{2.3}$$

As explained in Chap. 1, we will often use, in addition to b_i, the quantity $\bar{y}_i$ to define the line for row i.

The quantity $\bar{y}_i$ is the "height" corresponding to row i. We recall that $\bar{y}_i$ is the height (ordinate) of the fitted line at the value $\bar{x}$.

If we denote by $\hat{y}_{ij}$ the fitted value corresponding to y_{ij}, and by d_{ij} the residual from the linear-model fit, then

$$y_{ij} = a_i + b_i x_j + d_{ij}$$
$$= (\bar{y}_i - b_i \bar{x}) + b_i x_j + d_{ij}$$

or

$$y_{ij} = \bar{y}_i + b_i (x_j - \bar{x}) + d_{ij}. \tag{2.4}$$

We will continue the discussion of row-linear models in Chapter 4.

2.4 The Column-Linear Structure

If a set of two-way data is row-linear, then, in the *transpose* of the matrix of the data, the *columns* will be linearly related to each other, and consequently to the row averages. Such a structure is called *column-linear*.

For example, if we consider the transpose of the matrix in Table 2.3, we obtain Table 2.4. This set is columnlinear

Table 2.4. Column-Linearity - Transpose of Table 2.3

Reference

Wood, L.A. and G.M. Martin (1964). Compressibility of natural rubber at pressures below 500 kg/cm^2. *J. Res. Nat. Bur. Stand., Sec. A. Phys. Chem.*, 68A, 259–268.

3

Functional Expressions
of Structure

The structures we have encountered in Chapter 2 can be expressed in a very general way. It will be useful to be aware of these expressions.

3.1 Additive Structures

The additive structure is given by the general equation:

$$y_{ij} = A_i + B_j + \varepsilon_{ij}. \tag{3.1}$$

Here A_i is a function of the rows only, and B_j is a function of the columns only. An example will help to understand Eq. (3.1).

Consider a 3×4 matrix constructed as follows: We first choose *any* set of three numbers to represent the rows A_i, for example, 5, -2, 9. Then we choose *any* set of four numbers to represent the columns B_j, for example, -1, 3 -5, 6. Then we construct the additive table, for example, $y_{ij} = 2(A_i) + 10(B_j)$ (see Table 3.1).

Table 3.1. An Additive Structure

A_i \ B_j	-1	3	-5	6
5	0	40	-40	70
-2	-14	26	-54	56
9	8	48	-32	78

This is a strictly additive set. We can add a set of randomly selected errors, ε_{ij}, from any population of such errors to obtain a set represented by Eq. (3.1).

Instead of choosing arbitrary numbers, such as 5, -2, 9, ..., we could, of course, obtain functionally related numbers. For example, the rows could be obtained from the equation $F_{\text{rows}} = -4 + 3(x^2)$ and we could select $x = 1, 4,$ 10. Thus, $F_1 = -4 + 3$, $F_2 = -4 + 48$, and $F_3 = -4 + 300$.

The same technigue is used for the columns.

We see that A_i and B_j can be *any* function, linear or non linear.

3.2 Multiplicative Structures

We select a set of numbers for the rows and a second set of numbers for the columns, as before.

But this time we *multiply* the numbers by each other. Thus, choosing the same sets as before, we construct a table, say $y_{ij} = 2(A_i \cdot B_j)$ (see Table 3.2).

Table 3.2. A Multiplicative Structure

A_i \ B_j	-1	3	-5	6
5	-10	30	-50	60
-2	4	-12	20	-24
9	-18	54	-90	108

Thus, the equation of a multiplicative structure is

$$y_{ij} = A_i \cdot B_j + \varepsilon_{ij}. \tag{3.2}$$

(Note that the coefficient 2 in the equation $y_{ij} = 2(A_i B_j)$ can be absorbed either in A_i or B_j since $2A_i$ is still a function of rows only.)

3.3 Row-Linear Structure

The equation for this structure is

$$y_{ij} = A_i + B_i \cdot C_j + \varepsilon_{ij}. \tag{3.3}$$

The construction of a set satisfying this equation proceeds as before, but here we need *three* sets: A_i, B_i, and C_j. The first two are functions of the rows (i); and the third, C_j, is a function of the columns (j). For example,

$$
\begin{array}{lrrrr}
A_i: & 5 & -2 & 9 & \\
B_i: & 1 & 8 & -3 & \\
C_j: & -1 & 3 & -5 & 6
\end{array}
$$

A two-way table satisfying Eq. (3.3) but ignoring ε_{ij} would be that shown in Table 3.3.

Table 3.3. A Row–Linear Structure C_j

A_i	B_i	-1	3	-5	6
5	1	4	8	0	11
-2	8	-10	22	-42	46
9	-3	-6	-18	6	-27

If we now display the 3 tables *without* the numbers from which we derived them, as in Table 3.4, we see that to typify them as additive, multiplicative, and row-linear will require a bit of trial and error. We will however see that there are systematic ways of doing the identification.

Table 3.4. Tables 3.1, 3.2, and 3.3

0	40	-40	70	-10	30	-50	60	4	8	0	11
-14	26	-54	56	4	-12	20	-24	-10	22	-42	46
8	48	-32	78	-18	54	-90	108	-6	-18	6	-27

3.4 Effects of Coding

When dealing with very large numbers or very small numbers, it is often advantageous to *code* them prior to any further calculations. Coding consist in two operations:

a. Add to (or subtract from) all the numbers the same constant, say K.
b. Multiply (or divide) the resulting numbers by the same constant M, generally a power of 10.

For example, the set

$$1132.5 \qquad 1130.7 \qquad 1133.1 \qquad 1135.2$$

can be coded to

$$25 \qquad 7 \qquad 31 \qquad 52$$

by means of $K = -1130$, $M = 10$.

To *decode*, we need to know K and M and proceed in reverse order: divide by M, subtract K. Thus,

$$25/10 - (-1130) = 1132.5.$$

How does coding affect the structure of a two-way table?
We have the following:

(1) For the additive model,

$$K + M(A_i + B_j) = K + MA_i + MB_j$$
$$= (K + MA_i) + MB_j$$
$$= A_i^* + B_j^*.$$

An additive structure *remains* additive under coding.

(2) For the multiplicative structure,

$$K + M(A_i \cdot B_j) = K + M \cdot A_i \cdot B_j = K + (MA_i)B_j.$$

But this is a row-linear structure *of a special type*. Indeed we have $K + (M \cdot A_i) = A^* + B_i^* C_j^*$.

The special type arises from the fact that A^* is a constant rather than a function of i. We will have an opportunity to make closer acquaintance with this special structure.

Thus, *coding a multiplicative structure results in a row-linear structure of a special type.*

(3) For a row-linear structure,

$$K + M(A_i + B_i \cdot C_j = (K + MA_i) + (MB_i)C_j$$
$$= A_i^* + B_i^* \cdot C_j^*.$$

We see that *coding a row-linear structure results in another row-linear structure.*

3.5 Interrelationships Between Structures

It is easy to see that the structures we have studied are interrelated. Consider the row-linear structure

$$y_{ij} = A_i + B_i \cdot C_j + \varepsilon_{ij}.$$

If B_i is a costant, say K, then

$$y_{ij} = A_i + KC_j + \varepsilon_{ij}$$
$$= A_i^* + B_j^* + \varepsilon_{ij}.$$

Thus, the additive strutture is simly a *special case* of a row-linear structure in which B_i is the same for all rows. Similarly, if in the now-linear structue, we make $A_i = 0$ for all rows, then it beccomes obviously a multiplicative structure. Thus, the multiplicative structure, too, is a special case of the row-linear structure.

We conclude that the row-linear structure is the one we should study in detail, as the others are special cases of it.

3.6 Canonical Form of the Row-Linear Structure

It is useful to write the equation of the row-linear model [Eq. (3.3)] in a standardized form. We have

$$y_{ij} = A_i + B_i \cdot C_j + \varepsilon_{ij}. \tag{3.4}$$

Denoting the average of the y_{ij} in column j by x_j [1], and the average of the y_{ij} in row i by $\bar{y}_i$, we have

$$\bar{y}_i = A_i + B_i \cdot \bar{C} + \bar{\varepsilon}_i , \tag{3.5}$$

where ε_i is the average of the ε_{ij} in row i, and

$$x_j = \bar{A} + \bar{B} \cdot C_j + \bar{\varepsilon}_j , \tag{3.6}$$

where $\bar{\varepsilon}_j$ is the average of the ε_{ij} in column j. Averaging Eq. (3.6) over j yields

$$\bar{x} = \bar{A} + \bar{B} \cdot \bar{C} + \bar{\bar{\varepsilon}}, \tag{3.7}$$

where $\bar{\bar{\varepsilon}}$ is the grand average of the ε_{ij}. Equations (3.6) and (3.7) give us

$$x_j - \bar{x} = \bar{B} \cdot (C_j - \bar{C}) + (\bar{\varepsilon}_j - \bar{\bar{\varepsilon}}) \tag{3.8}$$

or

$$C_j = \bar{C} + \frac{(x_j - \bar{x}) - (\varepsilon_j - \bar{\bar{\varepsilon}})}{\bar{B}}. \tag{3.9}$$

From Equation (3.5) we obtain

$$A_i = \bar{y}_i - B_i \cdot \bar{C} - \bar{\varepsilon}_i . \tag{3.10}$$

Introducing Eq. (3.9) and (3.10) into Eq. (3.4) gives

$$y_{ij} = (\bar{y}_i - B_i \cdot \bar{C} - \bar{\varepsilon}_i) + B_i \cdot \left[\bar{C} + \frac{(x_j - \bar{x}) - (\bar{\varepsilon}_j - \bar{\bar{\varepsilon}})}{\bar{B}} \right] + \varepsilon_{ij}$$

or

$$y_{ij} = \bar{y}_i + \frac{B_i}{\bar{B}} \cdot (x_j - \bar{x}) + \left[(\varepsilon_{ij} - \bar{\varepsilon}_i) - \frac{B_i}{\bar{B}} \cdot (\bar{\varepsilon}_j - \bar{\bar{\varepsilon}}) \right]. \tag{3.11}$$

---

[1] We consider x_j as independent of y_{ij} and will justify this later.

The expression in brackets is the error term. Let us denote it by ε_{ij}^{*}.
Let us denote $B_i / \overline{B}$ by β and observe that $\overline{\beta} = 1$. Then Eq. (3.11) becomes

$$y_{ij} = \bar{y}_i + \beta_i (x_j - \bar{x}) + \varepsilon_{ij}^{*} \qquad (3.12)$$

The parameter β_i can be estimated by linear regression of y_{ij} on x_j for a fixed i. Denoting this estimate by b_i and noting that $\overline{b} = 1$, we obtain

$$y_{ij} = \bar{y}_i + b_i (x_j - \bar{x}) + d_{ij}, \qquad (3.13)$$

where d_{ij} is the residual of the regression. Equation (3.13) is the canonical form of the row-linear model.

4

The Row-Linear
and Concurrent Structures

We have already seen, in Chapter 2, how the row-linear model can be under-stood as consisting of a bundle of straight lines, one for each row, when the rows are plotted against their common column averages. We have also noted that the row-linear structure is often encountered in real scientific data and that the additive and multiplicative structures are special cases of it.

By its very definition, it is easy to ascertain whether a particular set of data obeys a row-linear structure. All we need to do is plot each row against the column averages and observe whether the graphs can be considered reasonable approximations to straight lines.

As an example, we show in Figure 4.1 a plot of the data from Table 2.1, which we displayed in Chapter 2 without discerning their structure. We now see that this structure is certainly very close to a row-linear one.

4.1 The Concurrent Structure

In addition to the additive and multiplicative structures, there is another special case for the row-linear structure that merits special attention; this is the structure that resulted from coding the multiplicative structure [case (2) in Section 3.4]:

$$y_{ij} = A + B_i \cdot C_j , \tag{4.1}$$

where A is a constant, independent of i.

Suppose that we find a j-value for which $C_j = O$. Then the corresponding y-value will be equal to A. But this is a constant independent of i. It follows that there exists a point, with coordinates $(C_j = 0, y = A)$ at which the straight lines corresponding to *all* of the rows meet. We will say that the structure in question is *concurrent*, as all the lines *concur* at one point.

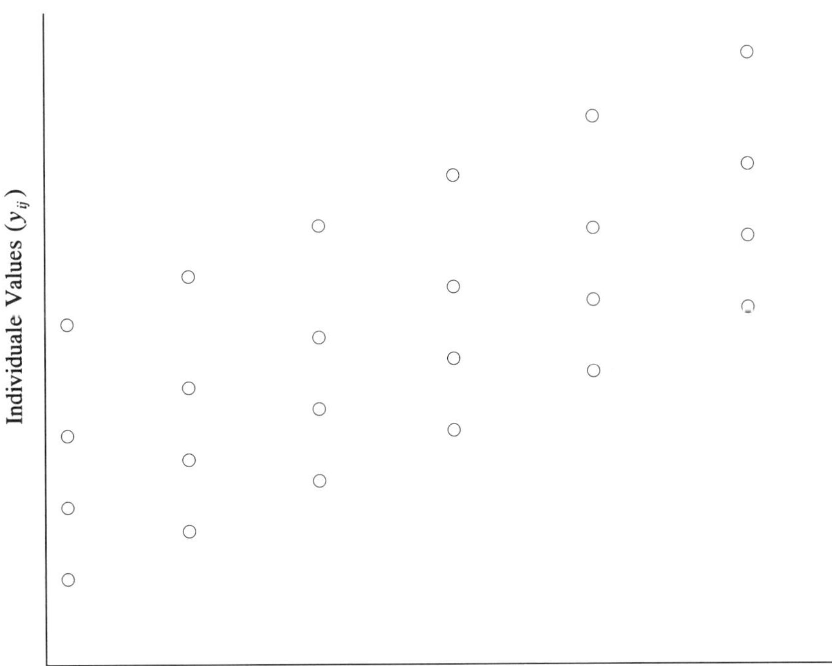

Figure 4.1. Specific volume of peroxide-cured rubber; y_{ij} versus x_j. Data of Table 2.1.

The concurrent structure is characterized by the equation

$$y_{ij} = A + B_i \cdot C_j + \varepsilon_{ij}. \tag{4.2}$$

It is also a special case of *row-linearity*.

Here again, the exact concurrent structure may be somewhat obscured by experimental errors ε_{ij}. The larger the ε_{ij}, the greater will be the departure from strict concurrence. A concurrent structure will result in a graph of y_{ij} versus C_j, in which the lines corresponding to the different rows will tend to fan out from a common point.

4.2 Characterization of a Row-Linear Structure: An Example

We consider once more Table 2.1. At the bottom of the table we have added a row, which we can label of column averages x_j. In Fig. 4.1 we exhibit the plot of all the rows versus x_j. The structure appears to row-linear, but because of the large range of values on the graph, it is not easy to see matters with sufficient detail. A simple remedy exists. Consider the equation

$$y_{ij} = A_i + B_i \cdot x_j + \varepsilon_{ij}.$$

Subtracting x_j from both sides gives

$$y_{ij} - x_j = A_i + (B_i - 1) \cdot x_j + \varepsilon_{ij}. \tag{4.3}$$

Equation (4.3) still represents a row-linear structure, but because x_j is the column-average and generally of the same order of magnitude as the y_{ij} for the same j- value, the quantity $y_{ij} - x_j$ will generally be small. The scale on the ordinate, $y_{ij} - x_j$, is now such that we will see much greater detail on the graph. A plot of $y_{ij} - x_j$ versus x_j, for all rows i, is shown in Fig. 4.2. In Table 4.1, we have listed the values resulting from fitting the straight lines y_{ij} versus x_j, the height $\bar{y}_i$ the slope b_i, the estimated standard deviation of ε, s_ε, and the standard deviation of the slope s_{b_i}. At the bottom of the table we also display the standard deviation of the four b_i values. It is important to understand the differences between this standard deviation, which we denote by $s(b)$, and the individual s_{b_i} values. The latter result from the fitting of each line, separately, against x_j. They result from the presence of ε_i. The value $s(b)$, on the other hand, measures the variability between the b_i values corresponding

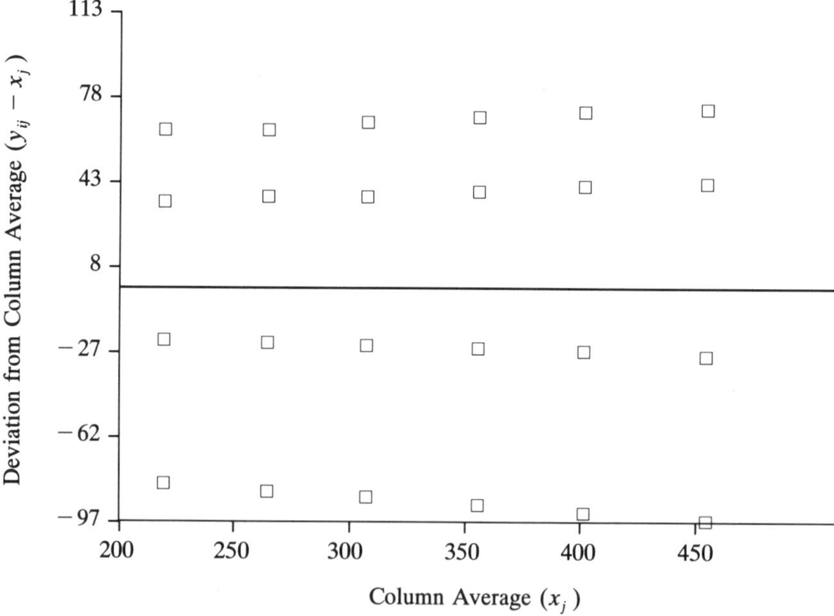

Figure 4.2. Specific volume of cured rubber; $y_{ij} - x_j$ versus x_j. Data of Table 2.1.

Table 4.1. Analysis of Table 2.1

Row	Height	Slope	s_ε	s_{b_i}
1	243.50	0.9356	0.51	0.0026
2	308.17	0.9821	0.29	0.0015
3	373.17	1.0310	0.26	0.0013
4	404.67	1.0512	0.44	0.0023
		s(b) = 0.0518		

to the four rows. We see that in our case, $s(b)$ is considerably larger than the individual s_{b_i} values. This means that the slopes of the four rows differ from each other more than can be accounted for by the ε_{ij} errors. In other words, the lines really do have different slopes. It follows that these data are definitely *not* additive.

The standard deviations, s_ε, listed in Table 4.1 are due entirely to the ε-values and are estimates of the standard deviation σ_ε for each row. We can also say that they are measures of the *residuals* of the fit of each row versus x_j. More information will be obtained if we display the residuals themselves.

Let us recall that the residual d_{ij} is given by

$$d_{ij} = y_{ij} - [\bar{y}_i + b_i \cdot (x_j - \bar{x})], \tag{4.4}$$

where the quantity in brackets is the fitted value $\hat{y}_{ij}$. Table 4.2 displays all the residuals, and Fig. 4.3 is a graph of the residuals.

Figure 4.2 displays a fanning-out of the lines corresponding to the various rows. This suggest that the data may be concurrent. There is a simple way to test for concurrence.

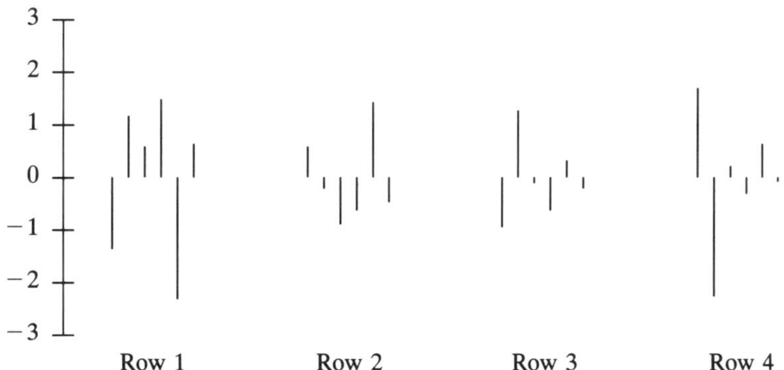

Figure 4.3. Specific volume of cured rubber; residuals[1] from row-linear fit. Data of Table 2.1.

[1] The residuals have been standardized by subtracting their mean and dividing by their standard deviation.

Table 4.2. Residuals from Row-Linear Fit. Data in Table 2.1

Row						
1	-0.423	0.345	0.177	0.436	-0.708	0.171
2	0.184	-0.048	-0.263	-0.178	0.434	-0.129
3	-0.278	0.389	0.026	-0.173	0.081	-0.046
4	0.517	-0.686	0.060	-0.086	0.193	0.003

4.3 A Test for Concurrence

We have seen that the basic relation for a row-linear structure, ignoring the experimental error, is

$$y_{ij} = A_i + B_i \cdot C_j \tag{4.5}$$

and that the condition for concurrence is

$$A_i = A \text{ (independent of } i). \tag{4.6}$$

Averaging Eq. (4.5) over j gives

$$\bar{y}_i = A_i + B_i \cdot \bar{C}. \tag{4.7}$$

Because of Eq. (4.6), Eq. (4.7) becomes

$$\bar{y}_i = A + B_i \cdot \bar{C}. \tag{4.8}$$

This is a linear relation between the height $\bar{y}_i$ and the slope B_i.

We have proved that such a relation is a *necessary* condition for concurrence.

We can readily prove that it is also a *sufficient* condition for concurrence. Indeed, if we neglect the error term in the equation for a row-linear structure and write

$$y_{ij} = A_i + B_i \cdot C_j , \tag{4.9}$$

where A_i, B_i, and C_j are population parameters, and if we have a linear relation between A_i and B_i such as

$$A_i = \alpha + \beta \cdot B_i , \tag{4.10}$$

then, introducing Eq. (4.10) into Eq. (4.9), we have

$$y_{ij} = (\alpha + \beta B_i) + B_i \cdot C_j$$

or

$$y_{ij} = \alpha + B_i \cdot (\beta + C_j),$$

where α is *not* a function of i. But this is how we have defined the concurrent structure because $\beta + C_j$ is a function of j only, and not of i.

In other words, in a row-linear structure, the existence of a linear relation between A_i and B_i (or between height and slope in the usual way of expressing the row-linear structure) is a necessary and sufficient condition for concurrence.

4.4 Point of Concurrence

If a set of data appears to be concurrent, it is of interest to determine the point in the (x, y) plane in which all of the lines concur. This is called the *point of concurrence.*

Consider the row-linear model

$$y_{ij} = \bar{y}_i + b_i(x_j - \bar{x}) + d_{ij} \qquad (4.11)$$

and the condition of concurrence (see Section 4.3)

$$b_i = P + Q\bar{y}_i , \qquad (4.12)$$

that is, a linear relation between $\bar{y}_i$ and b_i.

Averaging Eq. (4.12) over i, and remembering that $\bar{b} = 1$, we obtain

$$1 = P + Q\bar{x} \qquad (4.13)$$

because the average of all $\bar{y}_i$ is the grand average $\bar{x}$.

From Eq. (4.12) and (4.13) we obtain

$$b_i = (1 - Q\bar{x}) + Q\bar{y}_i$$

or

$$b_i = 1 + Q(\bar{y}_i - \bar{x}). \qquad (4.14)$$

Then Eq. (4.11) becomes

$$y_{ij} = \bar{y}_i + [1 + Q(\bar{y}_i - \bar{x})](x_j - \bar{x}) + d_{ij} . \qquad (4.15)$$

Let us define a quantity y_0 by

$$y_0 = \bar{x} - 1/Q \qquad (4.16)$$

and subtract Eq. (4.16) from Eq. (4.15):

$$y_{ij} - y_0 = (\bar{y}_i - \bar{x}) + 1/Q + (x_j - \bar{x}) + Q(\bar{y}_i - \bar{x}) \cdot (x_j - \bar{x}) + d_{ij} . \qquad (4.17)$$

Let us represent the quantities $\bar{y}_i - \bar{x}$ and $x_j - \bar{x}$ by:

$$U_i = (\bar{y}_i - \bar{x}) \quad \text{and} \quad V_j = (x_j - \bar{x}). \qquad (4.18)$$

Then (4.17) becomes

$$y_{ij} - y_0 = U_i + 1/Q + V_j + Q(U_i V_j) + d_{ij}$$

which can be written as

$$y_{ij} - y_0 = \frac{(1 + QU_i)(1 + QV_j)}{Q} + d_{ij}. \tag{4.19}$$

Now, from Eq. (4.16) we obtain

$$Q = \frac{1}{(\bar{x} - y_0)}. \tag{4.20}$$

Hence,

$$1 + QU_i = 1 + \frac{\bar{y}_i - \bar{x}}{\bar{x} - y_0} = \frac{\bar{y}_i - y_0}{\bar{x} - y_0}. \tag{4.21}$$

Similarly,

$$1 + QV_j = \frac{x_j - y_0}{\bar{x} - y_0}. \tag{4.22}$$

Because of Eqs. (4.21) and (4.22), we can write Eq. (4.19) as

$$y_{ij} - y_0 = \frac{(\bar{y}_i - y_0)(x_j - y_0)}{\bar{x} - y_0} + d_{ij}. \tag{4.23}$$

This equation shows that if $x_j = y_0$, then, except for experimental error, $y_{ij} = y_0$. Therefore, y_0 is the ordinate of the point of concurrence, and it is also its abscissa. We have shown that the point of concurrence has equal abscissa and ordinate. It is, therefore, on the bisector of the axes x and y, and its abscissa and ordinate are both equal to

$$y_0 = \bar{x} - 1/Q,$$

where Q is the slope of the straight line relating b_i to $\bar{y}_i$.

We note, finally, the important fact that a concurrent set of data is both row-linear and column-linear. This follows at once from Eq. (4.1).

4.5 Return to the Example

Now that we have a criterion for concurrence, let us apply it to our example (data of Table 2.1). Figure 4.4 is a plot of the slope against the height for the four rows of the table. There is a strong linear relation between the two, and we conclude that the structure of Table 2.1 is essentially concurrent. The point of concurrence is given by coordinates which we call x_0 and y_0:

$$x_0 = -1050, \qquad y_0 = -1050.$$

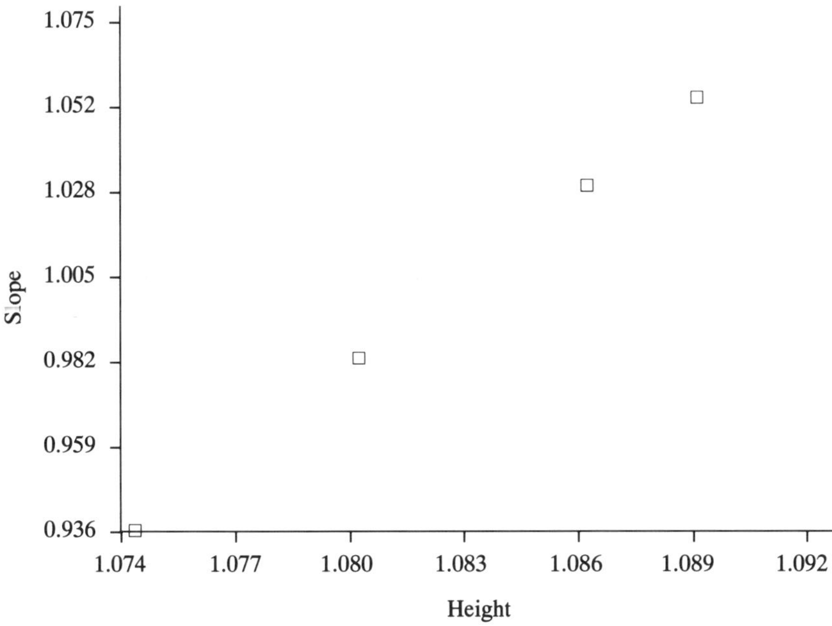

Figure 4.4. Slope versus height. Data of Table 2.1.

The point of concurrence is so far to the left of the graph that the lines are almost parallel. Nonetheless, our analysis has revealed non parallelism.

Our analysis of the data is far more instructive than the classical analysis of variance because the latter, through its pooling of many individual effects into the desired sums of squares, often hides many aspects that are of critical importance for an understanding of the data.

One point deserves to be discussed. Because of the experimental errors ε_{ij}, any structure, such as row-linearity or concurrence, is to some extent obscured. The larger the ε_{ij} the greater is the effect. As a consequence, one is seldom able to assign with certainty a particular structure to a table of real measurements. Often one will find that certain parts of the data obey a particular structure more fully than do others. The so-called "modeling" of data is subject to these considerations: A model is generally only a useful approximation, not an exact mathematical description of the data.

The analysis we have made of our numerical example is not only an approximation but it is also *incomplete* in the sense of allowing for other approaches that may throw further light on the data. This will be the subject of the next chapter.

4.6 The Interdependence of y_{ij} and x_j

Clearly, x_j is not strictly independent of y_{ij}, so that a regression of y_{ij} on x_j, using classical regression formulas, is not entirely valid. In this section we examine this problem in more detail.

In a matrix of p rows and q columns, it is, of course, possible to regress y_{ij} for a given i (a given row) on the average of all *other* rows. Let x'_{ij} represent the average of all elements in column j, omitting the one element y_{ij} for a given i. For example, if we consider the first row ($i = 1$), then $y_{ij} = y_{1j}$ and $x'_{ij} = x'_{1j} = $ the average of all elements, *except* the first, in column j. Obviously, we have

$$x'_{ij} = (px_j - y_{ij})/(p - 1) \qquad (4.24)$$

because px_j is the *sum* of all elements in column j.

If the variance of y_{ij} is σ^2, then the variance of x'_{ij} is $\sigma^2/(p - 1)$. Thus, y_{ij} and x'_{ij} are independent and the ratio of their variances is $\sigma^2/[\sigma^2/(p - 1)]$ $= p - 1$. We can then apply the results of Chapter 1 (both variables subject to error) with $\lambda = p - 1$.

Let us denote by b'_i the estimate of the slope obtained by regressing y_{ij} (for a fixed i) on x'_{ij}, considering both y_{ij} and x'_{ij} subject to error, with a lambda-value $\lambda = p - 1$. This is, of course, not the slope we wish to estimate. But we can obtain the latter in the following way.

We write the two equations, regression of y_{ij} on x_j and regression of y_{ij} on x'_{ij}:

$$y_{ij} = a_i + b_i x_j , \qquad (4.25)$$

$$y_{ij} = a'_i + b'_i x'_{ij} \qquad (4.26)$$

Using Eq. (4.24), we have

$$y_{ij} = a'_i + b'_i \left[\frac{px_j - y_{ij}}{p - 1} \right]$$

or

$$y_{ij} \left[1 + \frac{b'_i}{p - 1} \right] = a'_i + b'_i \left[\frac{p}{p - 1} \right] x_j$$

and, consequently,

$$y_{ij} = \frac{a'_i}{1 + \dfrac{b'_i}{(p - 1)}} + \frac{b'_i [p/(p - 1)]}{1 + \dfrac{b'_i}{(p - 1)}} x_j .$$

Thus,

$$b_i = \frac{pb_i'}{p - 1 + b_i'}. \tag{4.27}$$

This equation allows us to estimate b_i from b_i' and circumvent the lack of independence of x_i and y_{ij}. The standard deviation, say s_b^*, can also be derived from this equation. Indeed, the derivative of b_i on b_i' is

$$\frac{d(b_i)}{d(b_i')} = \frac{p(p - 1 + b_i') - 1(pb_i')}{(p - 1 + b_i')^2} = \frac{p(p - 1)}{(p - 1 + b_i')^2},$$

so that

$$s_{b_i}^* = \frac{p(p - 1)}{(p - 1 + b_i')^2} \cdot s_{b_i} \tag{4.28}$$

The question we now have to address is how good the approximation of b_i (obtained by classical regression of y_{ij} on x_j) is, as judged from its closeness to b_i derived as above, and also how close s_b (derived from classical regression on x_j) is to $s_{b_i}^*$.

For large p, we would expect the approximation (of regressing y_{ij} on the nonindependent x_j) to be good, as many rows are involved in calculating x_j. But for small p, the question is open and should be examined. Table 4.3 is an artificial set of data with a fairly large error term. We have made the regression analysis both ways: using x_j and using x_j', and then converting to the proper slope and standard deviation of the slope [Eqs. (4.27) and (4.28)]. The results are shown in Table 4.4.

Table 4.3. An Artificial Set of Data

Row				
1	16.5	2.6	1.8	8.2
2	34.3	14.0	8.7	20.7
3	9.2	3.8	4.5	8.4

Table 4.4. Results for Data of Table 4.3

Row	Height	Slope		Std. Error of Slope	
		Regression x_j	Regression x_j'	Regression x_j	Regression x_j'
1	7.275	1.0011	1.0011	0.0540	0.0540
2	19.425	1.6303	1.6382	0.1216	0.1219
3	6.475	0.3685	0.3615	0.1154	0.1154

We see that the results are almost identical. Thus, even for three rows, the approximation for the regression on x_j is excellent. For larger numbers of rows, the results are indistinguishable from each other, as shown by a number of examples we used to test the approximations.

We conclude that we need not worry about the nonindependence of y_{ij} and x_j, and can use the regression on x_j with confidence.

5

Graphical Representation of the Row-Linear Structure

5.1 Introduction

In the previous chapters we have discussed a number of structures often encountered in two-way arrays of data. We have also presented some graphical procedures for the display of data of this type. However, two-way arrays lend themselves to other interesting and useful graphical analyses. In this chapter we present two types of graphs that are highly instructive when trying to understand two-way data.

5.2 The *h* Graph and the *h'* Graph

We denote, as before, the (i, j) entry of the table by y_{ij}, the average of all elements in column j by $\bar{x}_j$ and the average of all elements in row i by $\bar{y}_i$. The grand average is denoted by $\bar{x}$. We do not assume any particular structure for the data and we note that the procedures we are about to explain are totally general and require no such assumptions.

We require two additional quantities, which we denote by $(s_r)_i$ and $(s_c)_j$. They are defined as follows:

$$(s_r)_i = \sqrt{\Sigma_j (y_{ij} - \bar{y}_i)^2/(q - 1)}, \tag{5.1}$$

$$(s_c)_j = \sqrt{\Sigma_i (y_{ij} - \bar{x}_j)^2/(p - 1)}. \tag{5.2}$$

Obviously, $(s_r)_i$ is the estimated standard deviation among elements in row i, and $(s_c)_j$ is the estimated standard deviation among elements in column j. We now define

$$h_{ij} = (y_{ij} - \bar{x}_j)/(s_c)_j, \tag{5.3}$$

$$h'_{ij} = (y_{ij} - \bar{y}_i)/(s_r)_i. \tag{5.4}$$

The statistics h_{ij} and h'_{ij}; are seen to be essentially "z-scores" of the jth column and the ith row, respectively. Note that h_{ij} is calculated for all elements of column j, *without making use of any other column of the table*. Similarly, h'_{ij} is calculated for all elements in row i *without* making use of any other row of the table.

We use h mainly for the purpose of comparing a particular cell average with other cell averages in the same column. It can be shown that for this end use, the statistical distribution of h' can be derived from that of a Student's-t variate with $(p - 2)$ degrees of freedom. Similar considerations apply to h'.

The computation of all h_{ij} and h'_{ij} is readily computerized and is extremely rapid on the computer. We note the following properties of h_{ij} and h'_{ij}:

$$\Sigma_i (h_{ij})^2 = \frac{\Sigma_i (y_{ij} - \bar{x}_j)^2}{(s_c)_j^2} = \frac{(p - 1)(s_c)_y^2}{(s_c)_j^2} = p - 1, \qquad (5.5)$$

$$\Sigma_j (h'_{ij})^2 = \frac{\Sigma_j (y_{ij} - \bar{y}_i)^2}{(s_r)_i^2} = \frac{(q - 1)(s_r)_i^2}{(s_r)_i^2} = q - 1. \qquad (5.6)$$

Therefore, we have

$$-\sqrt{p - 1} \leq h_{ij} \leq +\sqrt{p - 1}, \qquad (5.7)$$

$$-\sqrt{q - 1} \leq h'_{ij} \leq +\sqrt{q - 1}. \qquad (5.8)$$

From Eqs. (5.3) and (5.4) it easily follows that h_{ij} and h'_{ij} can be positive or negative. A positive value of h_{ij} indicates a value of y_{ij} *above* the column average $\bar{x}_j$; a negative value of h_{ij} indicates a value of y_{ij} *below* the column average of $\bar{x}_j$. Similar statements can be made for h'_{ij} in terms of y_{ij} being *above* or *below* $\bar{y}_i$.

It is important to realize that h'_{ij} is simply the h_{ij} of the *transposed* matrix of y_{ij} data. Finally, we observe the following important facts.

The h_{ij} calculated for y_{ij} values in the same *column* of the table are *not* statistically independent, but they preserve the *relative* positions of the y_{ij} (of that particular column) with respect to each other. The h'_{ij} calculated for y_{ij} values in the same *row* of the table are *not* statistically independent, but they preserve the *relative* positions of the y_{ij} (of that particular row) with respect to each other.

We will now illustrate the h_{ij} and h'_{ij} values for a particular set of data, and then discuss their graphs and their interpretation.

5.3 A Numerical Example

Table 5.1 is taken from an article relating to two-way classifications (Johnson and Graybill, 1972). It displays the yield of spring wheat as a function of nitrogen and phosphorus content of the soil.

Table 5.1. Yield of Spring Wheat (kg/ha)

Nitrogen (kg/ha)	Phosphorus (kg/ha)				
	0	22	45	90	180
0	1984	2550	2706	2740	2954
45	1776	2843	3306	3305	3386
90	1797	2761	3240	3227	3332

Table 5.2. h-Values for Table 5.1

1.150	-1.111	-1.149	-1.145	-1.147
-0.667	0.827	0.675	0.700	0.688
-0.483	0.284	0.474	0.445	0.459

Table 5.3. h'-Values for Table 5.1

-1.645	-0.100	0.325	0.418	1.002
-1.697	-0.119	0.566	0.565	0.684
-1.677	-0.172	0.575	0.555	0.719

The h-values are displayed in Table 5.2. and Fig. 5.1, and the h'-values are displayed in Table 5.3 and Fig. 5.2. Note that the h are assembled on the graphs in groups representing *rows* of the table, and the h' in groups representing *columns* of the table.

We will first demonstrate an important fact about h and h' that can help us in the interpretation of the plots of these quantities.

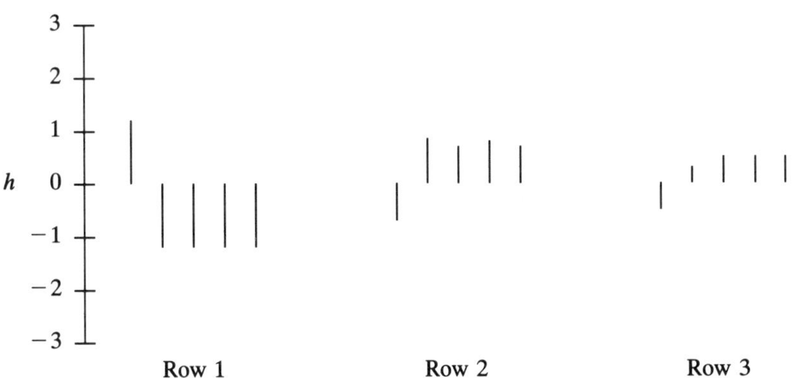

Figure 5.1. h-Values for data of Table 5.1.

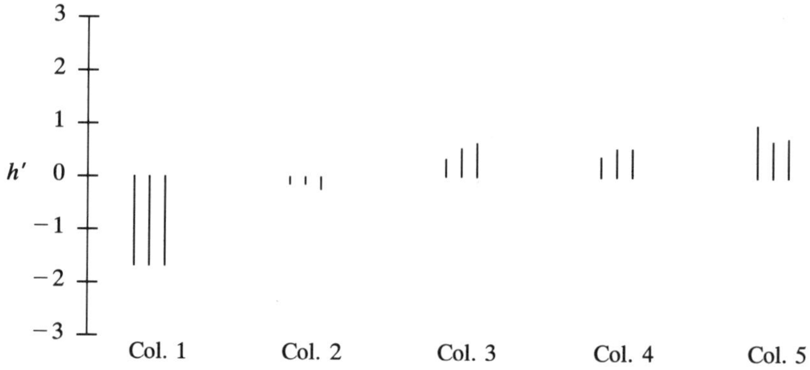

Figure 5.2. h'-Values for data of Table 5.1.

5.4 An Important Property of h and h'

Suppose that a particular set of data is row-linear, and consider the h' statistic. We have [see Eq. (5.2)]

$$(s_c)_j = \sqrt{\Sigma_i (y_{ij} - \bar{y}_i)^2 / (p-1)} \, .$$

If the set is row-linear, we have

$$y_{ij\cdot} - \bar{y}_i = b_i(x_j - \bar{x}) + d_{ij} \, .$$

It follows that

$$\Sigma_j (y_{ij} - \bar{y}_i)^2 = b_i^2 \Sigma_j (x_j - \bar{x})^2 + \text{error terms.}$$

Consequently

$$(h'_{ij})^2 = \frac{(y_{ij} - \bar{y}_i)^2}{(s_r)_i^2}$$

$$= \frac{b_i^2(x_j - \bar{x})^2 + \text{error terms}}{[b_i^2 \Sigma_j (x_j - \bar{x})^2 + \text{error terms}]/(q-1)} \, .$$

If we neglect the error terms, we obtain the approximate relation:

$$(h'_{ij})^2 \approx \frac{(x_j - \bar{x})^2}{\Sigma_j (x_j - \bar{x})^2} (q-1) \, . \tag{5.9}$$

Because the right-hand member of Eq. (5.9) is *independent* of i, it follows that, approximately, all of the $(h'_{ij})^2$ for any particular column j have the same magnitude. Therefore, all of the h'_{ij} in any particular column, except possibly for sign, will be approximately equal to each other.

We have proved that for a row-linear model, all h'_{ij} in any column have *approximately* the same absolute value.

It is important to observe that the converse is not necessarily true. If the h' in every group have essentially the same absolute value, and if this is true for all groups, then it does *not* follow that the matrix is necessarily row-linear. The reason for that is if the column effect is very large with respect to experimental error, the same behavior pattern will be observed, even if the data are not row-linear. We will discuss this in greater detail by means of an artificial example in Section 5.6.

If this property is observed to hold in an h' plot, then this may be (but is not necessarily) a clue as to row-linearity of the matrix .

Similarly, if we observe that in an h plot, all h_{ij} for the same row have essentially the same absolute value, and this holds for all rows, then we may suspect (but have not proved) column-linearity of the original matrix.

5.5 Return to the Example

The h and h' plots for the data in Table 5.1 *both* display, to a good approximation, the equality of absolute values in any particular row and column, respectively. We, therefore, may suspect the data to be approximately both row-linear and column-linear, hence concurrent. The exact test of this statement requires some insight into the results of an analysis of variance, a topic to be dealt with in the next chapter.

Even though h and h' cannot determine with finality the structure of a set of data, they are useful in many ways. They have the merit of showing us the behavior of individual rows and columns. Thus, it is clear from Fig. 5.2 that column 1 is very different from (much lower than) the other four columns and that, among the latter, column 2 is somewhat lower than columns 3, 4, and 5. Indubitably, h and h' are an excellent way of obtaining an instant picture of the general pattern of the data. We will see later that they are a powerful tool in the analysis of interlaboratory data.

5.6 Discussion of an Artificial Example

Table 5.4 is a set of artificial data, introduced to obtain a better understanding of the h and h' graphs.

The h and h' graphs are shown in Fig. 5.3 and 5.4. From our discussion in Section 5.4 we would be tempted to conclude that, because of the appearance of h', they are row-linear. We will see later that the data are column-linear but *not* row-linear. Note that the range of values, over columns, is quite large, and

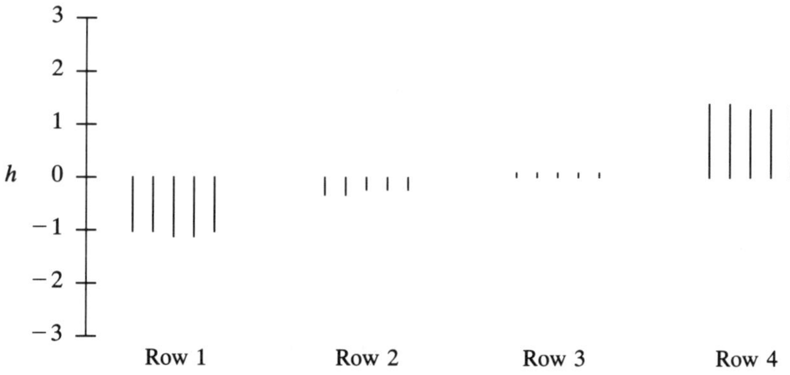

Figure 5.3. *h*-Values for data of Table 5.4.

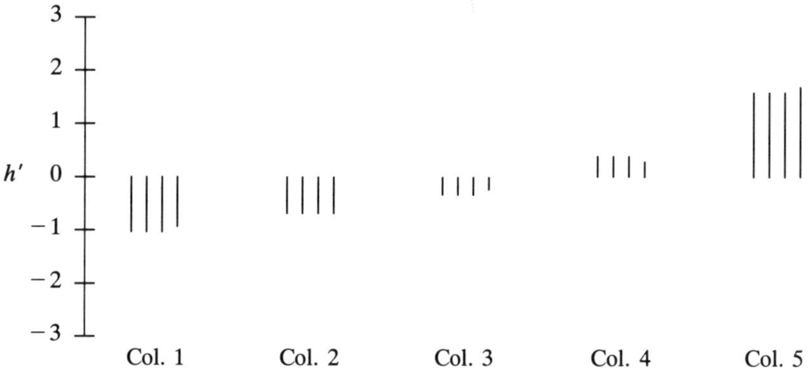

Figure 5.4. *h'*-Values for data of Table 5.4.

Table 5.4. An Artificial Example

1.22	8.91	16.48	30.67	56.30
3.55	10.68	19.50	32.11	58.92
4.79	11.60	20.98	32.79	60.17
8.42	14.31	25.48	34.90	64.11

this has caused h' to appear as it does even though the data are not row-linear. We will return to this example.

References

Johnson, D.E. and F.A. Graybill (1972). Estimation of σ^2 in a two-way classification model with interaction. *J. Amer. Statist. Assoc.* 67, 388–394.

6

Some Analysis of Variance

6.1 Introduction

Continuing the discussion of the example (Table 5.1) used in the previous chapter, we note that we have essentially five possibilities:

1. the data are row-linear
2. the data are column-linear
3. the data are concurrent
4. the data are additive
5. none of the above properly describes the structure of the data

Our purpose in this chapter is to present a numerical technique for choosing among these possibilities.

6.2 A General Principle

Let us assign symbols to the five possibilities above: ROW-L for row-linear, COL-L for column-linear, CONC for concurrent, AD for additive, and NONE for possibility 5.

Schematically, we have the following arrangement:

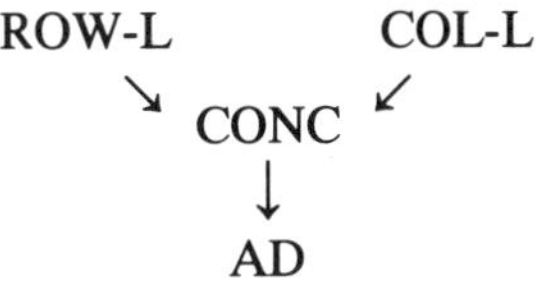

In this scheme, the first row contains independent possibilities. The second row is a special case of both structures in the first row. The third row is a special case of the second row.

Thus, AD is a special case of CONC, and CONC is a special case of both ROW-L and COL-L.

We now state that for two structures, for which one is a special case of the other, we can judge the advantage of one over the other by examining the residuals for each of the two fits. Table 6.1 list the sums of squares of residuals, the corresponding degrees of freedom, and the mean squares for the four cases, for the data of Table 5.1.

Table 6.1. Analysis of Variance of Data of Table 5.1

Structure	Sum of Squares of Residuals	Degrees of Freedom	Mean Square
ROW–L	1,389	4	347
COL–L	22,494	6	3,749
CONC	22,880	7	3,268
AD	258,652	8	32,332

It can be shown that the sum of squares of residuals for a "special case" structure is never smaller, and generally larger, than the sum of squares of residuals for the more general structure. Therefore, the adoption of a more specific structure is generally obtained at the price of *increasing* the sum of squares of residuals. The question then arises: Is it reasonable to pay this price?

6.3 Further Discussion of Table 5.1

Consider rows 3 and 4 in Table 6.1. Denote by SSR the sum of squares of residuals. Adopting AD over CONC increases SSR from 22,880 to 258,652. Is AD a reasonable model for these data?

The answer given by statistical theory is as follows: If G and S represent the general and specific model, then consider the ratio F (where DF represents degrees of freedom):

$$F = \frac{(\text{SSR}_S - \text{SSR}_G/(\text{DF}_S - \text{DF}_G)}{\text{SSR}_G/\text{DF}_G}. \tag{6.1}$$

This is an "F-ratio with (DF_S - DF_G) and DF_G degrees of freedom." The critical values of F are tabulated in the work of Pearson and Hartley (1966). For our example, we have

$$F = \frac{(258{,}652 - 22{,}880)/(8 - 7)}{22{,}880/7} = 72.13.$$

This ratio has 1 and 7 degrees of freedom. It is very highly significant, and we conclude that the additive model is not satisfactory for these data. The theory, when applied to CONC versus ROW-L, gives an F-ratio of 20.63, with 3 and 4 degrees of freedom. This also is significant at the 1% level and indicates that the concurrent model itself is not entirely satisfactory for our data.

The statistical literature advocates, to a large degree, a "test for non-additivity" devised by J.W. Tukey (1949), which is a comparison of (AD − CONC) with CONC.

This is precisely the test we used to evaluate AD. However, we used it to evaluate additivity in comparison with concurrence only. Our conclusion relates to the superiority, for these data, of the concurrent model over the additive model. We did *not* conclude that concurrence was a "good" model. In fact, we showed that it was *not* satisfactory, by comparing it with the row-linear model, because it increased significantly the sum of squares of residuals.

The statistical theory we have just used does *not* lend itself to the comparison of structures that are not "embedded" one in the other. Thus, it does not give us the means to compare directly, for example, ROW-L and COL L. Of course, in our case (Table 6.1), we hardly need a statistical test: The residuals for ROW-L are so much smaller than those from COL-L that there is no doubt that the former is a much better fit than the latter.

It is important to realize that, here as always in using analysis of variance, it is essential to first ascertain that there are no substantial outliers in the data. Indeed, the presence of outliers could completely invalidate the comparison of sums of squares of residuals simply because even a single outlier can increase this sum of squares very substantially.

6.4 Further Discussion of Table 5.4

If we consider the sums of squares of residuals for the data of Table 5.4 we obtain the results shown in Table 6.2.

Table 6.2. Analysis of Variance of Data of Table 5.4

Structure	Sum of Residuals of Squares	Degrees of Freedom	Mean Square
ROW–L	7.5074	9	0.8342
COL–L	0.00577	8	0.000722
CONC	7.5102	11	0.6827
AD	7.5850	12	0.6321

The superiority of COL-L over ROW-L requires no test. It is overwhelming. Testing CONC versus COL-L yields an F-value of over 3000, so we confidently eliminate CONC as a probability. Note that this conclusion is obtained in spite of the appearance of the h' graph which would cause us to consider row-linearity as a possibility. Of course, the mean squares in Table 6.2 are such that no statistical test is necessary to obtain the proper conclusion. The standard deviation of the error term is $\sqrt{0.000722} = 0.027$, a very small quantity when evaluated in terms of the range of values exhibited in Table 5.4.

References

Pearson, E.S. and H.O. Hartley (1966). *Biometrika Tables for Statisticians*, Cambridge, Vol.1, Cambridge University Press.

Tukey, J.W. (1949). One degree of freedom for non-additivity. *Biometrics*, 5, 232–242.

7

A General Representation of Two-Way Arrays

7.1 Introduction

So far, we have demonstrated the usefulness of a number of techniques for the elucidation of two-way tables of measurements. First, there is the plotting of each row versus the column averages, or each column versus the row averages. A more sensitive procedure, in general, is a plot of the *deviation* of the measurement from the column average versus column average, or a similar plot for *deviations* from the row average versus row average.

A very useful procedure, applicable to any two-way table without any assumptions as to model, is the plotting of h in groups of rows and h' in groups of columns. The analysis of variance, comparing sums of squares of residuals, is useful in judging whether a more specific model can be used to replace a more general model in which it is embedded. This technique should always be accompanied by a graphical look at the individual residuals, since a sum of squares can be very misleading because of its inherent pooling characteristic.

In this chapter we introduce a general technique applicable to any two-way array of numbers. We will present this technique, as well as an algorithm that implements it on the computer. We will then attempt to interpret the results and examine whether they can be interpreted in terms of specific structures such as row-linearity, column-linearity, or concurrence. Although this technique generally requires the availability of a computer (without it, the calculations would be very tedious), the computer algorithm itself is very simple. We highly recommend this technique to anyone who has access to a computer.

7.2 A Single Multiplicative Term

Consider a set of data, for example Table 5.1, and suppose that we attempt to fit the data by the formula

$$y_{ij} = A_i \cdot B_j + d_{ij}. \tag{7.1}$$

In this equation, A_i is a function of rows only, and B_j is a function of columns only. The residual, d_{ij}, is what is left after fitting $A_i \cdot B_j$. It is advantageous to write the multiplicative term, $A_i \cdot B_j$, in a slightly different form:

$$A_i \cdot B_j = \theta \cdot u_i \cdot v_j \tag{7.2}$$

Here u_i and v_j are standardized vectors such that

$$\Sigma_i (u_i)^2 = 1 \quad \text{and} \quad \Sigma_j (v_j)^2 = 1, \tag{7.3}$$

and θ is a positive constant. u_i is obtained by the formula

$$u_i = \frac{A_i}{\sqrt{\Sigma_i (A_i)^2}} \tag{7.4}$$

and v_j by

$$v_j = \frac{B_j}{\sqrt{\Sigma_j (B_j)^2}}. \tag{7.5}$$

Therefore,

$$u_i v_j = \frac{A_i B_j}{\sqrt{\Sigma_i (A_i)^2 \Sigma_j (B_j)^2}} \tag{7.6}$$

and

$$\theta = \sqrt{\Sigma_i (A_i)^2 \Sigma_j (B_j)^2}. \tag{7.7}$$

As an example, let

$$A_i = [-3, 7] \qquad (p = 2),$$
$$B_j = [9, 7, 6, 1] \qquad (q = 4).$$

Then

$$u_1 = -3/\sqrt{(-3)^2 (7)^2} = -0.3939,$$
$$u_2 = 7/\sqrt{(-3)^2 (7)^2} = 0.9191$$

and, similarly,

$$v = [0.6964 \quad 0.5417 \quad 0.4643 \quad 0.0774]$$

For θ, we find

$$\theta = \sqrt{(58)(167)} = 98.4175.$$

The form $\theta u_i v_j$ for the term $A_i B_j$ allows us to assess at once the magnitude of the term. Indeed, since both u_i and v_j lie between -1 and $+1$, θ is a measure of the magnitude of the term.

The problem we have to solve is to find A_i and B_j (and consequently θ, u_i, and v_j), given a matrix y_{ij}. The solution is based on the idea that the multiplicative term must be chosen so as to make d_{ij} as small as possible. This can be done by applying the principle of least squares, which leads to the following equations:

$$\theta \cdot u_i = \Sigma_j (y_{ij} \cdot v_j) \tag{7.8}$$

$$\theta \cdot v_j = \Sigma_i (y_{ij} \cdot u_i). \tag{7.9}$$

To obtain θ, u_i, and v_j from these equations is not a straightforward problem. A number of algorithms are available. We will present a simple algorithm and illustrate it with data of Table 5.1. However, the data in this have been coded by dividing each value by 1000.

7.3 Algorithm for Finding θ, u_i and v_j

As shown in Table 7.1 we place a vector of three numbers (because $p = 3$) to the right of the table. We chose the numbers 1, 2, and 3, but any set of numbers (not all the same) can be used (Step 1).

Table 7.1. Algorithm for SVD. Data of Table 5.1 (Divided by 1000)

	Data					(1)	(4)	(5)	(8)	(9)
	1.984	2.550	2.706	2.740	2.954	1	5.8168	0.5283	5.8205	0.5286
	1.776	2.843	3.306	3.305	3.386	2	6.6740	0.6061	6.6725	0.6060
	1.797	2.761	3.240	3.227	3.332	3	6.5469	0.5946	6.5459	0.5945
(2)	10.927	16.519	19.038	19.031	19.722		$\downarrow$			
(3)	0.2816	0.4257	0.4906	0.4904	0.5082		$\theta = 11.0114$			
(6)	3.1931	4.7120	5.3599	5.3695	5.5941					
(7)	0.2900	0.4279	0.4868	0.4876	0.5080					

We next calculate successively the *inner product* of this set with each column of the table. An inner product of two vectors of equal length is the sum of the cross-products of corresponding elements. For example, the inner product of the vector [1, 2, 3] with the first column of Table 7.1 is

$$(1 \times 1.984) + (2 \times 1.776) + (3 \times 1.797) = 10.927.$$

An inner product of two vectors is a number, not a vector. We place this number under column 1 and repeat the process for the other columns, as shown in Table 7.1 (Step 2).

This vector is now *standardized*, as follows: Calculate the square root of the sum of squares of the numbers and divide each number of the vector by this square root (Step 3).

Next, we form successively, the inner product of this new row with, the three rows of the table and place the results to the right of the table (Step 4). This new vector is standardized (Step 5).

The entire process is now repeated, using this standardized vector, instead of the vector [1, 2, 3]. After a few iterations, the standardized vector on the right will no longer change. This will be the u_i vector.

Using this u_i, we then obtain a row vector formed from the inner product of u_i with each column of the table. The standardized form of this vector is v_j, and θ is the square root of the sum of squares of the elements of this vector prior to its standardization. Having obtained θ, u_i and v_j, we attempt to reconstruct the table by calculating for each cell the quantity

$$\hat{y}_{ij} = \theta \cdot u_i \cdot v_j . \tag{7.10}$$

Then,

$$d_{ij} = y_{ij} - \hat{y}_{ij} . \tag{7.11}$$

7.4 The Complete Singular Value Decomposition

The entire process described in Section 7.3 can now be applied to the matrix d_{ij}. In this fashion, we can write

$$y_{ij} = \theta_1 \cdot u_{1i} \cdot v_{1j} + \theta_2 \cdot u_{2i} \cdot v_{2j} + \cdots , \tag{7.12}$$

where each term is obtained from the residuals left after fitting all previous terms.

Equation (7.12) is called the *Singular Value Decomposition* (SVD) of the original y_{ij} matrix, and the constants θ_1, θ_2, and so on are called the Singular Values. The square of a singular value, θ^2, is called an *eigenvalue*, or *latent root*. An important aspect of the SVD is that it *always* terminates. By that we mean that on the right-hand side of Eq. (7.12) the number of multiplicative terms $(\theta_k \cdot u_{ki} \cdot v_{kj})$ is always *finite*. Afler that, θ becomes zero. The number of nonzero θ-values is called the *rank* of the matrix y_{ij}, and it is a fact that after the number of terms equal to the rank have been obtained, the residuals are zero, so that Eq. (7.12), with all the nonzero terms, is *exact* (not an approximation). Our algorithm will always yield θ-values in nonincreasing order ($\theta_1 \geq \theta_2 \geq \theta_3 \geq \cdots$).

Furthermore, the rank of the matrix is equal to or smaller than the number of rows or the number of columns of the table, whichever is smaller. Thus, the rank of Table 5.1 is, at most, 3; three multiplicative terms, at most, in Eq. (7.12) will produce the matrix y_{ij} exactly. Table 7.2 presents the complete SVD of Table 5.1. We now ask, What is the purpose of making an SVD of a matrix?

Table 7.2. SVD of Table 5.1 (Divided by 1000)

$$
\begin{aligned}
\theta_1 &= \quad 11.0114 \\
u_1 &= [\quad 0.5286 \qquad 0.6060 \qquad 0.5945\,] \\
v_1 &= [\quad 0.2900 \qquad 0.4279 \qquad 0.4868 \qquad\qquad 0.4876 \qquad 0.5080\,] \\[6pt]
\theta_2 &= [\quad 0.4044 \\
u_2 &= [-0.8460 \qquad 0.4354 \qquad 0.3078\,] \\
v_2 &= [-0.8676 \quad -0.1679 \qquad 0.3694 \qquad\qquad 0.2873 \qquad 0.0070\,] \\[6pt]
\theta_3 &= [\quad 0.0304 \\
u_3 &= [-0.0723 \quad -0.6656 \qquad 0.7428\,] \\
v_3 &= [\quad 0.3020 \quad -0.8530 \qquad 0.3433 \qquad -0.0336 \qquad 0.2494\,]
\end{aligned}
$$

7.5 Approximations Made Possible by Singular Value Decomposition

By its very nature, SVD yields a set of θ-values, θ_1, θ_2, θ_3, that never increase. Thus,

$$\theta_1 \geq \theta_2 \geq \theta_3 \geq \cdots. \tag{7.13}$$

In addition, θ_1 will be as large as possible in comparison to d_{ij}, θ_2 will be as large as possible in comparison to the second set of residuals, and so en. Therefore, the first term, $\theta_1 \cdot u_{1i} \cdot v_{1j}$, will attempt to be as close as possible to the original matrix. But $\theta_1 \cdot u_{1i} \cdot v_{1j}$ is a term that separates rows (i) from columns (j) and will, therefore, lend itself to an examination of y_{ij} in terms of rows and columns. A powerfiul tool to do this is the *biplot*.

7.6 The Biplot

The biplot is a procedure due to Gabriel and Bradu (1978). Suppose that a matrix y_{ij} can be satisfactorily approximated by two terms,

$$y_{ij} = \theta_1 \cdot u_{1i} \cdot v_{1j} + \theta_2 \cdot u_{2i} \cdot v_{2j} + d_{ij}, \tag{7.14}$$

where d_{ij} represents experimental error, and suppose that a linear relation exists between v_{1j} and v_{2j}; say

$$v_{2j} = \alpha + \beta \cdot v_{1j}. \tag{7.15}$$

Introducing Eq. (7.15) into Eq. (7.14) gives

$$y_{ij} = \theta_1 \cdot u_{1i} \cdot v_{1j} + \theta_2 \cdot u_{2i} \cdot (\alpha + \beta \cdot v_{1j}) + d_{ij}$$

or

$$y_{ij} = \alpha \cdot \theta_2 \cdot u_{2i} + (\theta_1 \cdot u_{1i} + \beta \cdot \theta_2 \cdot u_{2i}) \cdot v_{1j} + d_{ij}.$$

But this is the form

$$y_{ij} = A_i + B_i \cdot C_j + d_{ij}$$

which represents a row-linear structure.

Thus, if, for Eq. (7.14), a linear relation exist between v_{1j} and v_{2j}, then y_{ij} has a *row-linear* structure. Similarly, if, in Eq. (7.14), a linear relation exists between u_{1i} and u_{2i}, then y_{ij} has a *column-linear* structure. Gabriel and Bradu suggest a plot of u_{2i} versus u_{1i} and of v_{2j} versus v_{1j} which they call a *biplot*. They call the (u_{1i}, u_{2i}) points the row markers and the (v_{1i}, v_{2i}) points the *column markers*. Linearity of the rowmakers indicates a column-linear model, and linearity in the columnmarkers indicates a row-linear model.

Figure 7.1 is a biplot of the data in Table 5.1. We see that both sets tend to be linear, the rowmarkers much more so than the columnmarkers. We conclude that Table 5.1 is essentially column-linear, and only very roughly row-linear.

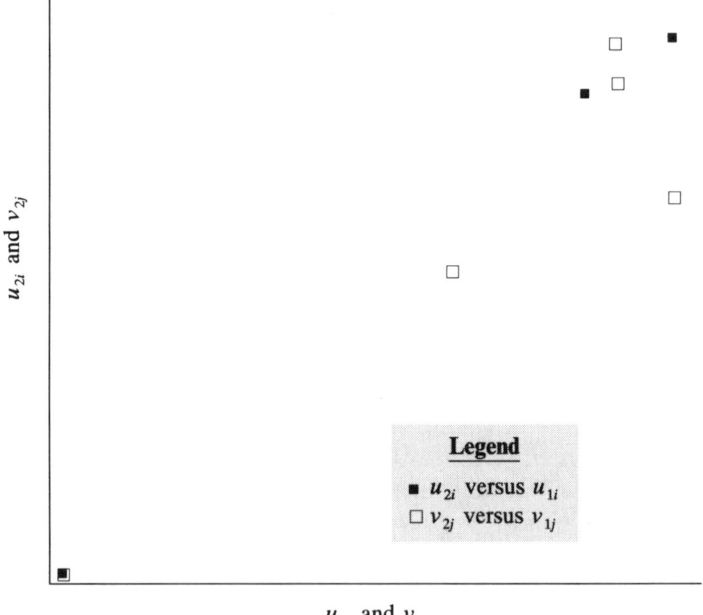

Figure 7.1. Biplot for Table 5.1.

7.7 Additive and Multiplicative Terms

A procedure analogous to the biplot, but providing generally greater sensitivity, has been developed by the author (Mandel 1971). Consider a matrix y_{ij}. Calculate the row averages $\bar{y}_i$, the column averages x_j, and the grand average $\bar{x}$. We then write

$$y_{ij} = \bar{x} + (\bar{y}_i - \bar{x}) + (x_j - \bar{x}) + d_{ij}^* \qquad (7.16)$$

This is always possible, and $\bar{x}$, $(\bar{y}_i - \bar{x})$, and $(x_j - \bar{x})$ are called the *additive terms*. $(\bar{y}_i - \bar{x})$ is called the *row effect* and $(x_j - \bar{x})$ is called the *column effect*. If the residual d_{ij}^*, is merely experimental error, then y_{ij} is additive. We propose to make a SVD, *not on y_{ij} but on d_{ij}^**:

$$d_{ij}^* = \theta_1 \cdot u_{1i} \cdot v_{1j} + \theta_2 \cdot u_{2i} \cdot v_{2j} + \cdots . \qquad (7.17)$$

Let us suppose that only one term is required to obtain a satisfactory approximation. Then we can write, omitting the index 1 in θ_1, u_{1i}, and v_{1j},

$$y_{ij} = [\bar{x} + (\bar{y}_i - \bar{x}) + (x_j - \bar{x})] + [\theta \cdot u_i \cdot v_j] + d_{ij} . \qquad (7.18)$$

The first bracket contains the additive term, the second the multiplicative terms, and d_{ij} is a residual of this dual fit.

We first note that the averages of $(\bar{y}_i - \bar{x})$ and $(x_j - \bar{x})$ are both zero. It can also be shown that in Eq. (7.17) the averages of u_i and v_j are also zero.

Let us now suppose that a linear relation exists between v_j and $(x_j - \bar{x})$:

$$v_j = \beta \cdot (x_j - \bar{x}). \qquad (7.19)$$

No intercept is required because v_j and $(x_j - \bar{x})$ both have zero average. Then Eq. (7.18) becomes

$$y_{ij} = \bar{x} + (\bar{y}_i - \bar{x}) + (x_j - \bar{x}) + \theta \cdot \beta \cdot u_i \cdot (x_j - \bar{x}) + d_{ij}$$

or

$$y_{ij} = \bar{y}_i + (1 + \theta \cdot \beta \cdot u_i) \cdot (x_j - \bar{x}) + d_{ij} . \qquad (7.20)$$

This again is of the form

$$y_{ij} = A_i + B_i \cdot C_j + d_{ij}$$

which is row-linear. This suggests that we plot v_j versus $(x_j - \bar{x})$ or, more simply, versus x_j. Similarly, we plot u_i versus $\bar{y}_i$. We can refer to these points as, column markers and row markers, respectively.

The plot of u_i versus $\bar{y}_i$ and of v_j versus x_j is similar to the biplot, but generally is more sensitive than the latter. Figure 7.2 exhibits this plot for Table 5.1.

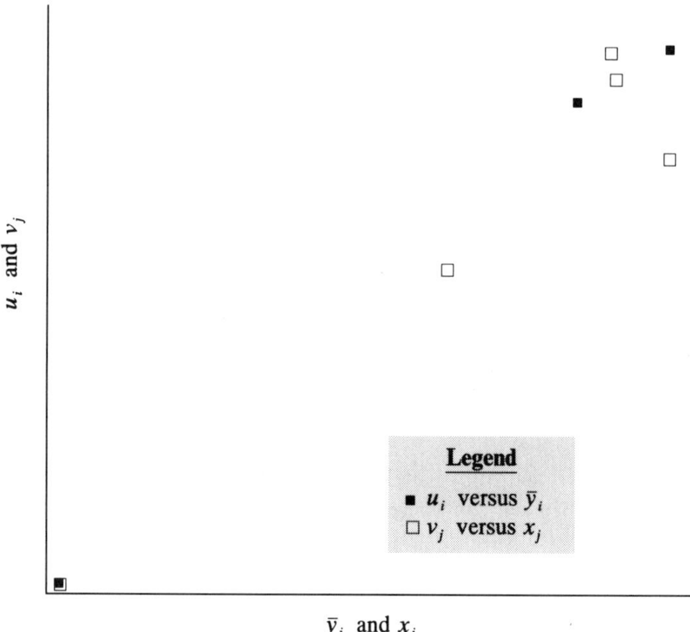

Figure 7.2. Biplot of Table 5.1 using AMA model.

We shall refer to this procedure as AMA, for Additive–Multiplicative-Analysis, and to Fig. 7.2 as a *diagnostic* plot.

Once we have ascertained that the data are free of serious outliers, AMA is a very sensitive indicator of row-linearity or column-linearity. We will see that it can also be very useful in the analysis of two-way tables with quantitative labels.

It is worth pointing out that the validity of AMA depends on the validity of Eq. (7.18), that is, on the fact that the SVD of the residuals (d_{ij}^*), after taking out the additive terms, requires only a single term. This condition should be checked by extending the SVD to more terms (at least one more) and ascertaining that the second term is small enough to be considered as experimental error. Let us observe that we have for the θ terms in the SVD of d_{ij}^* the identity

$$\Sigma_i \Sigma_j d_{ij}^{*2} = \theta_1^2 + \theta_2^2 + \cdots + \theta_r^2, \tag{7.21}$$

where θ_1^2 is the last nonzero θ-value (r is the rank of d_{ij}^*); let us observe that we have

$$\theta_1^2 \geq \theta_2^2 \geq \cdots \geq \theta_r^2 \tag{7.22}$$

so that, if θ_2^2 is very small compared to θ_1^2, then one term, $\theta_1 \cdot u_{1i} \cdot v_{1j}$, is, enough to represent the d_{ij}^* terms.

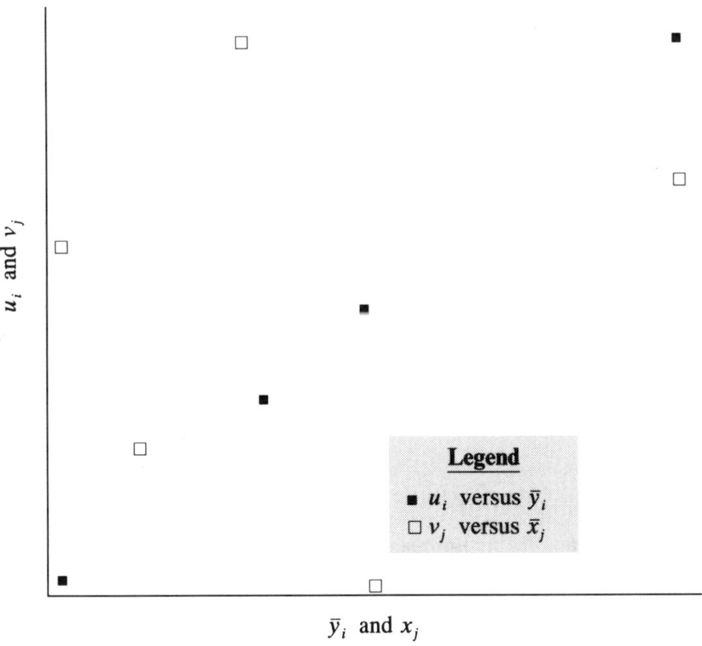

$\bar{y}_i$ and x_j

Figure 7.3. Diagnostic plot of Table 5.4 using AMA model.

For the d_{ij}^{*} of Table 7.1, we have

$$\theta_1 = 11.01139 \quad \text{and} \quad \theta_2 = 0.4044 \,.$$

For the data of Table 5.4, the diagnostic plot is shown in Fig. 7.3. The first two θ-values in the SVD of the residuals are

$$\theta_1 = 2.7532 \quad \text{and} \quad \theta_2 = 0.0684 \,.$$

References

Gabriel, K.R. and Dan Bradu (1978). The biplot as a diagnostic tool for models of two-way tables. *Technometrics* 20, 47–68.

Mandel, John (1971). A new analysis of variance model for non-additive data. *Technometrics* 13, 1–18.

8

A Systematic Approach to the Analysis of Two-Way Arrays

8.1 Introduction

We now know enough techniques to attempt a systematic approach to the analysis of a set of two-way data. It is assumed that the analyst has access to a computer. Programming the various techniques on the computer is not difficult, as all techniques, with the possible exception of SVD, are computationally very simple. Using the algorithm for SVD explained in Chapter 7, even SVD is readily programmed.

An important point is that we never attempt to force the data into any specific model. On the contrary, what we attempt to do is to find, on the basis of a detailed examination of the data, a model that might represent them. It is by no means certain that such a model always exists. In many situations we will find that some models will represent the data but only in an approximate way. In such cases one must decide whether the approximation is close enough to make the model practically useful. This depends, of course, on the intended end use of the data and can be determined only by the subject matter specialist, not by the statistician.

8.2 Statistical Procedures

We illustrate the systematic approach first in terms of the data in Table 8.1, which represents a collaborative study of the curing characteristics of SBR rubber (Stiehler et. al., 1966), by means of the property t_2.

A good way is to start with h and h' plots for the data (Fig. 8.1 and 8.2). In general, these will reveal, among other things, whether severe outliers are present. If one finds outliers, the subject-matter specialist must decide what to do about them. This decision should include an examination of the probable

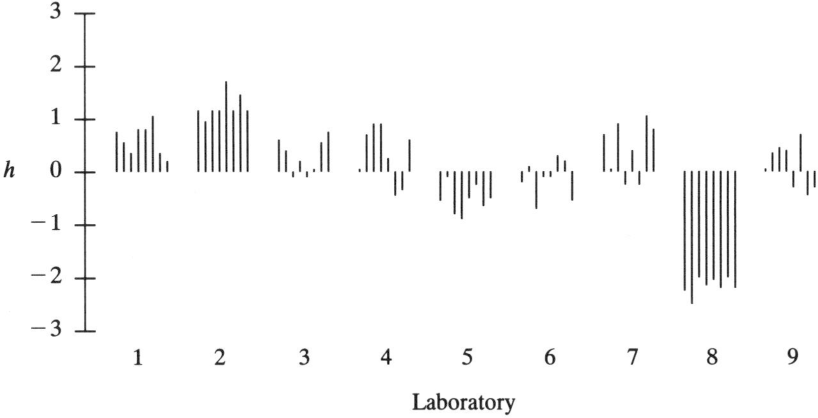

Figure 8.1. Graph of h-values. Data of Table 8.1.

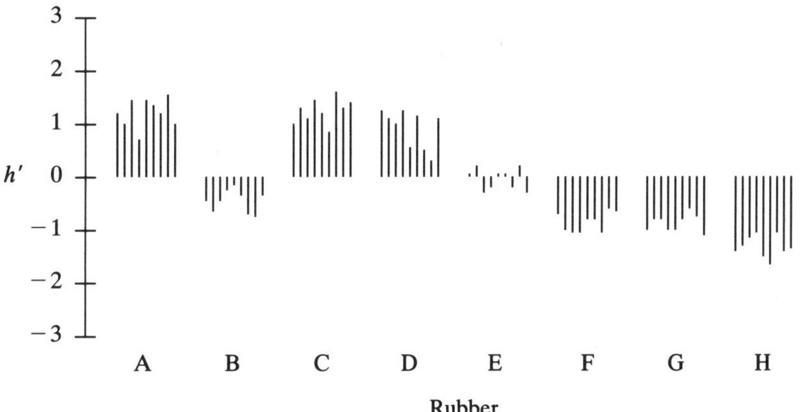

Figure 8.2. Graph of h'-values. Data of Table 8.1.

causes for the outliers. Often this leads to better insight into the experimental procedure or procedures and may result in substantial improvements. It should be noted that the widespread practice of *rejecting* outliers on the basis of statistical tests is philosophically indefensible and practically harmful. After such rejection, the data certainly "look better," but the causes that may lead to bad data have not been investigated, and, therefore, the usefulness of the data is open to question.

For our example, an examination of Fig. 8.1 shows that there is, indeed, a question as to whether *all* of the data are acceptable. Laboratory 8 is strikingly low. Because we cannot take the action recommended above (look for causes),

Table 8.1. Rheometric Cure of SBR Rubber. t_2 (min)

Lab	Rubber							
	A	B	C	D	E	F	G	H
1	16.80	12.80	16.50	17.00	13.90	12.10	11.50	10.40
2	17.50	13.20	18.40	17.90	15.20	12.20	12.80	11.40
3	16.60	12.60	15.80	15.60	12.80	11.20	11.80	11.00
4	15.60	13.00	17.80	17.20	13.20	10.80	10.90	10.80
5	14.70	12.10	14.30	13.20	12.20	11.00	10.60	9.80
6	15.40	12.30	14.40	15.00	12.80	11.40	11.40	9.80
7	16.80	12.20	17.80	14.80	13.40	11.00	12.30	11.00
8	11.80	9.00	11.50	10.20	10.00	9.20	9.00	8.10
9	15.60	12.60	16.80	16.00	12.50	11.80	10.80	10.00

we will ignore this disturbing feature and proceed with the analysis of *all* the data. Figure 8.2, which presents h' for these data leads us to examine whether they are row-linear.

In general, h and h' can provide hints at row- and column-linearity but, by themselves, do not prove the validity of such models. The next step is to make an AMA analysis. The graph of u_i versus $\bar{y}_i$ and of v_j versus x_j is shown in Fig. 8.3 (data of Table 8.1). The graph shows evidence of both row- and column-linearity. Next, we examine the residuals for various models by analysis of variance, as shown in Table 8.2.

Table 8.2. Analysis of Variance of Table 8.1

Structure	Sum of Squares of Residuals	Degrees of Freedom	Mean Square
ROW-L	16.8220	48	0.3505
COL-L	16.4407	49	0.3355
CONC	18.9886	55	0.3452
AD	32.6389	56	0.5828

The increase in the sum of squares of residuals, when going from CONC to AD, is highly significant and we conclude that an additive model is out of the question. A test for the concurrent structure against either the row linear or column-linear structure shows that a concurrent structure is compatible with the data.

The theory of the row- and column-linear models allows us to look at one more aspect of the data. Figure 8.4 is a plot of the slope versus height, in the fitting of a row-linear model. There is a strong linear relationship between height and slope, confirming the appropriateness of a concurrent fit (See Section 4.3). But of special interest in this plot is the position of the "outlying

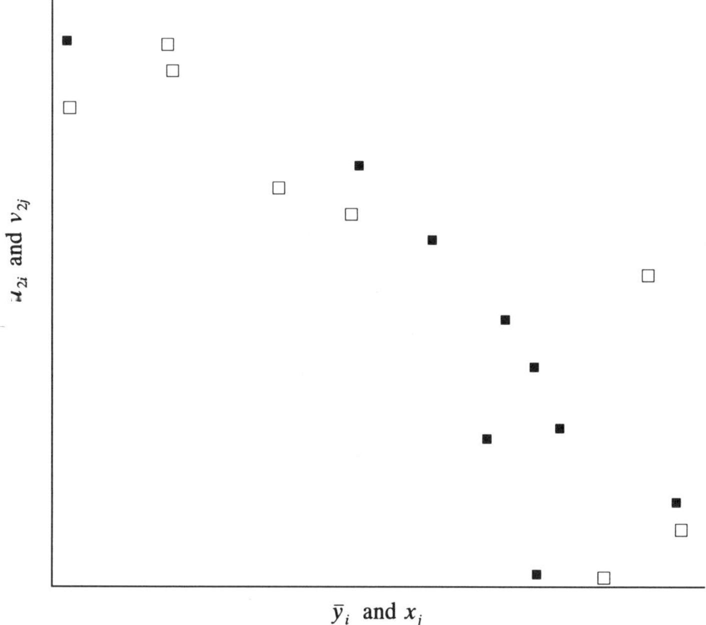

Figure 8.3. Diagnostic plot of Table 8.1.

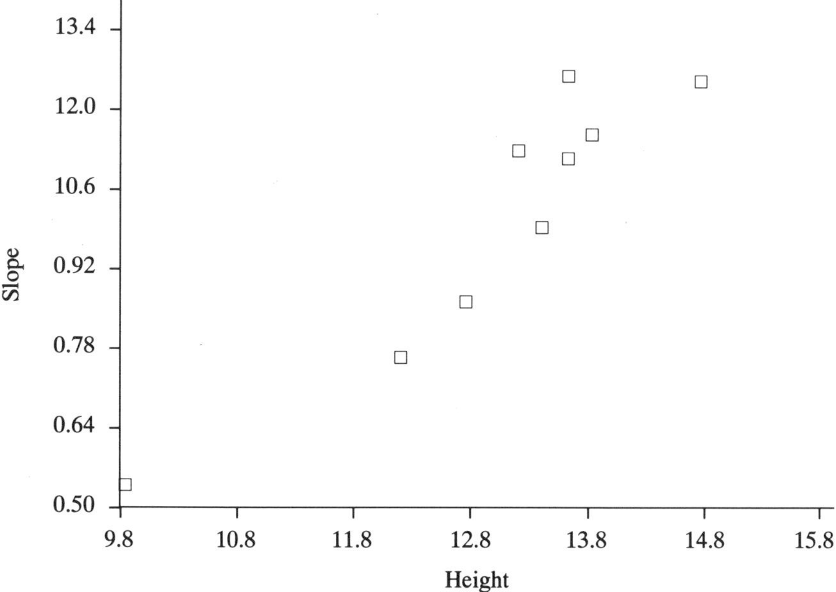

Figure 8.4. Slope versus height (Table 8.1).

laboratory" (laboratory 8). We see that the point representing this laboratory fits rather well on the straight line defined by the other points. Thus, although laboratory 8 is very low as compared to the other laboratories, it has a slope consistent with the low values. Its position as an outlier must be qualified: Its values are not haphazardly wrong; rather they prove some systematic departure from the other laboratories, perhaps an erroneous calibration. It would have been interesting, and probably fruitful, to ascertain the cause of the discrepant values obtained by this laboratory.

8.3 Further Discussion of Table 5.4

We return to our discussion of the data in Table 5.4 (an artificial set). We have already looked at the h and h' graphs for these data (Section 5.6), and we have looked at the analysis of variance of residuals for them (Section 6.4). From the latter analysis we have concluded that the data are essentially column-linear. In Section 7.7, we have looked at the AMA diagnostic plot for the data (Fig. 7.3). This plot, more than any other analysis, clearly shows the data to be column-linear and *not* row-linear. Indeed, the row markers in Fig. 7.4 are quite linear, but the column-markers fail completely a linearity test.

In reflecting upon the various analyses we have made, we see that the h and h' plots, although extremely useful for obtaining a general picture of the pattern of the data, are not sufficient by themselves to ascertain row- or column-linearity. The analysis of the variance of residuals is useful for the *comparison* of four models (row-linear, column-linear, concurrent, or additive). It also provides measures for the error standard deviation (e.g., $0.59 = \sqrt{0.3452}$ for the Cure data, adopting a concurrent model). The AMA plot is quite convincing, showing us the plausibility of a column-linear structure for the data of Table 5.4 and the non admissibility of a row-linear structure.

Clearly, it pays to look at the data from all of these viewpoints, and we strongly recommend that this be done as a general operational procedure for the analysis (or rather, *internal* analysis) of any set of two-way data.

Of course, following this course of action is often insufficient: When any puzzling situation is revealed in any of these approaches, action is required. By action we mean efforts to *explain* the anomalies through questioning of, and conferring with, the persons who made the measurements. Simply *rejecting* the data one does not like is, as pointed out before, an illogical course of action; one that sweeps the discrepancies under the rug instead of confronting them. We strongly oppose rejection of data on the basis of purely statistical considerations.

References

Stiehler, R.D., M.N. Steel, J. Mandel, and G.E. Decker (1966). SBR Interlaboratory Test, NBS Report 9328.

9

Two-Way Analysis
with Replication Within Cells

9.1 Introduction

So far, we have dealt with two-way tables in which, for each row and column combination, there was a single measurement. This case is called "two-way arrays with single measurement per cell." It happens often, however, that several "replicate" measurements are found in each cell. For example, in interlaboratoly studies of a test method, it is useful to "replicate within cells." Consider Table 9.1 (NBS data). Here there are 18 rows, representing 18 laboratories. Each of these measured five materials (serums). The five serums contained increasing amounts of beta-carotene and were analyzed for this constituent. Thus, the table contains 18 times $5 = 90$ cells. But each material was analyzed by each laboratory in duplicate. Thus, the number of replicates per cell was 2 in this case.

We will assume that the table, in general, has p rows and q columns (and, therefore, pq cells), and n replicates in each cell. If there is only one measurement per cell, then $n = 1$.

9.2 Analysis of Replicated Two-Way Arrays

The analysis of such tables consists of two parts. First, we average the n replicates in each cell and analyze the resulting table (Table 9.2) of cell averages by the methods described in the preceding chapters (see also Chapter 10). Then we construct a new table of standard deviations within cells, denoted s_{ij}, as shown in Table 9.3.

For example, the first entry in Table 9.3 reads 0.0028, which is the standard deviation of the duplicates 0.066 and 0.062. Note that if $n = 2$, the standard deviation can be calculated directly as $d/\sqrt{2}$, where d is the absolute value of the difference between the duplicates. Thus, $|0.062 - 0.066|/\sqrt{2} = 0.0028$.

Table 9.1. Beta-carotene in Serum (μg/ml)

Lab.	A	B	C	D	E
1	0.066	0.146	0.472	0.986	1.708
	0.062	0.143	0.436	0.904	1.564
2	0.070	0.140	0.390	0.840	1.250
	0.070	0.140	0.390	0.820	1.240
3	0.089	0.213	0.496	1.286	1.897
	0.082	0.196	0.523	1.241	1.915
4	0.044	0.120	0.452	1.292	1.844
	0.050	0.120	0.472	1.131	1.824
5	0.064	0.142	0.411	0.883	1.372
	0.058	0.148	0.416	0.874	1.347
6	0.076	0.149	0.399	0.886	1.204
	0.073	0.145	0.396	0.859	1.288
7	0.080	0.230	0.390	0.830	1.160
	0.080	0.250	0.380	0.780	1.160
8	0.062	0.140	0.370	0.890	1.330
	0.057	0.150	0.390	0.910	1.400
9	0.060	0.170	0.450	1.040	1.520
	0.070	0.170	0.460	1.070	1.500
10	0.071	0.155	0.458	1.093	1.436
	0.074	0.159	0.444	1.061	1.555
11	0.050	0.140	0.420	0.970	1.630
	0.060	0.140	0.420	0.980	1.480
12	0.060	0.080	0.180	0.320	0.470
	0.050	0.060	0.190	0.350	0.510
13	0.051	0.145	0.371	0.832	1.260
	0.062	0.145	0.328	0.870	1.210
14	0.100	0.240	0.520	1.380	1.820
	0.090	0.230	0.590	1.180	2.040
15	0.063	0.146	0.426	0.899	1.405
	0.060	0.149	0.458	1.002	1.414
16	0.095	0.173	0.437	0.969	1.554
	0.097	0.177	0.447	0.978	1.571
17	0.070	0.138	0.389	0.914	1.408
	0.069	0.149	0.393	0.919	1.402
18	0.040	0.090	0.230	0.540	0.550
	0.040	0.090	0.230	0.530	0.530
Average over 18 laboratories	0.067	0.153	0.406	0.925	1.382

Table 9.2. Table of Cell Averages (Data of Table 9.1)

	0.0640	0.1445	0.4540	0.9450	1.6360
	0.0700	0.1400	0.3900	0.8300	1.2450
	0.0855	0.2045	0.5095	1.2635	1.9060
	0.0470	0.1200	0.4620	1.2115	1.8340
	0.0610	0.1450	0.4135	0.8785	1.3595
	0.0745	0.1470	0.3975	0.8725	1.2460
	0.0800	0.2400	0.3850	0.8050	1.1600
	0.0595	0.1450	0.3800	0.9000	1.3650
	0.0650	0.1700	0.4550	1.0550	1.5100
	0.0725	0.1570	0.4510	1.0770	1.4955
	0.0550	0.1400	0.4200	0.9750	1.5550
	0 0550	0.0700	0.1850	0.3350	0.4900
	0.0565	0.1450	0.3495	0.8510	1.2350
	0.0950	0.2350	0.5550	1.2800	1.9300
	0.0615	0.1475	0.4420	0.9505	1.4095
	0.0960	0.1750	0.4420	0.9735	1.5625
	0.0695	0.1435	0.3910	0.9165	1.4050
	0.0400	0.0900	0.2300	0.5350	0.5400
Averages	0.0671	0.1533	0.4062	0.9253	1.3824

Table 9.3. Table of Cell Standard Deviations (Data of Table 9.1)

	0.0028	0.0021	0.0255	0.0580	0.1018
	0.0000	0.0000	0.0000	0.0141	0.0071
	0.0049	0.0120	0.0191	0.0318	0.0127
	0.0042	0.0000	0.0141	0.1138	0.0141
	0.0042	0.0042	0.0035	0.0064	0.0177
	0.0021	0.0028	0.0021	0.0191	0.0594
	0.0000	0.0141	0.0071	0.0354	0.0000
	0.0035	0.0071	0.0141	0.0141	0.0495
	0.0071	0.0000	0.0071	0.0212	0.0141
	0.0021	0.0028	0.0099	0.0226	0.0841
	0.0071	0.0000	0.0000	0.0071	0.1061
	0.0071	0.0141	0.0071	0.0212	0.0283
	0.0078	0.0000	0.0304	0.0269	0.0354
	0.0071	0.0071	0.0495	0.1414	0.1556
	0.0021	0.0021	0.0226	0.0728	0.0064
	0.0014	0.0028	0.0071	0.0064	0.0120
	0.0007	0.0078	0.0028	0.0035	0.0042
	0.0000	0.0000	0.0000	0.0071	0 0141
Pooled std. deviation	0.0045	0.0065	0.0177	0.0511	0.0588

A standard deviation of two numbers has one degree of freedom and is highly unreliable. Nevertheless, we calculate the values in Table 9.3 because in their totality (i.e., for all 90 cells) they can provide us with useful information. This will be discussed in the following section.

9.3 The *k* Statistic

The s_{ij} occurring in any column of Table 9.3 represent elements of a single population if all laboratories work with the same "repeatability error." Repeatability refers to the ability of a laboratory to obtain similar (close) values if it repeats measurements on the same sample. In the case in which repeatability is the same in all laboratories, the s_{ij} in column j, each of which has $(n - 1)$ degrees of freedom, can be *pooled* to yield a pooled value, denoted $(s_r)_j$, of the standard deviation of replication error in column j.

The pooled value is obtained by

$$(s_r)_j = \sqrt{(s_{1j}^2 + s_{2j}^2 + \cdots + s_{pj}^2/p}\,. \tag{9.1}$$

The pooled value has $p(n - 1)$ degrees of freedom. In Table 9.3, the pooled values are shown at the bottom of the table. Note that they increase from left to right, indicating that as the beta-carotene in a sample increases, so does the standard deviation of replication error.

The pooling of standard deviations is based on the assumption that the pooled elements all come from the same population. We need to test this assumption and do it graphically by means of the k statistic. This statistic is defined by

$$k_{ij} = s_{ij}/(s_r)_j\,. \tag{9.2}$$

Thus, k is the ratio of the s_{ij} for cell (i, j) to the pooled value for that same column, j. The k-values for Table 9.3 are exhibited in Table 9.4.

The k-values listed in the same column of the original table are not statistically independent, but they preserve the relative positions of the various laboratories (in terms of repeatability) with respect to each other. Therefore, it is interesting to assemble them row-wise, as shown in Fig. 9.1. Here, the 5 k-values for each laboratory are assembled, and this is done for all 18 rows. The interpretation of this figure is straight forward. We see, for example, that all k-vatues for laboratory 14 are large, indicating that laboratory 14 showed consistently poorer repeatability than all the other laboratories. This invalidates, of course, the pooling over laboratories but does not invalidate our analysis, the purpose of which was precisely to exhibit the data in detail.

Table 9.4. Table of *k* Values (Data of Table 9.1)

0.635	0.327	1.440	1.135	1.731
0.000	0.000	0.000	0.277	0.120
1.111	1.855	1.080	0.623	0.216
0.952	0.000	0.800	2.228	0.240
0.952	0.655	0.200	0.125	0.301
0.476	0.436	0.120	0.374	1.010
0.000	2.182	0.400	0.692	0.000
0.793	1.091	0.800	0.277	0.842
1.587	0.000	0.400	0.415	0.240
0.476	0.436	0.560	0.443	1.431
1.587	0.000	0.000	0.138	1.804
1.587	2.182	0.400	0.415	0.481
1.745	0.000	1.720	0.526	0.601
1.587	1.091	2.801	2.768	2.645
0.476	0.327	1.280	1.426	0.108
0.317	0.436	0.400	0.125	0.204
0.159	1.200	0.160	0.069	0.072
0.000	0.000	0.000	0.138	0.240

It is tempting to "reject" laboratory 14, but, as pointed out previously, such rejection has no useful effect. In fact, it is misleading because it "improves" the data without improving the method of test. If any action is to be taken, it is an examination of the *causes* for the apparent aberration of laboratory 14. Such a course of action may well lead to an improvement of the analytical procedure for the determination of beta-carotene in serum. Our analysis is simply a way to call attention to anomalies, such as shown by the *k*-values for laboratory 14. It is also possible that the committee that conducts the inter-laboratory study decides that the aberration of laboratoly 14 in repeatability can be tolerated and that no further action is necessary. The rationale for such possible action will be discussed further in Chapter 10, as part of the overall analysis of interlaboratoly data.

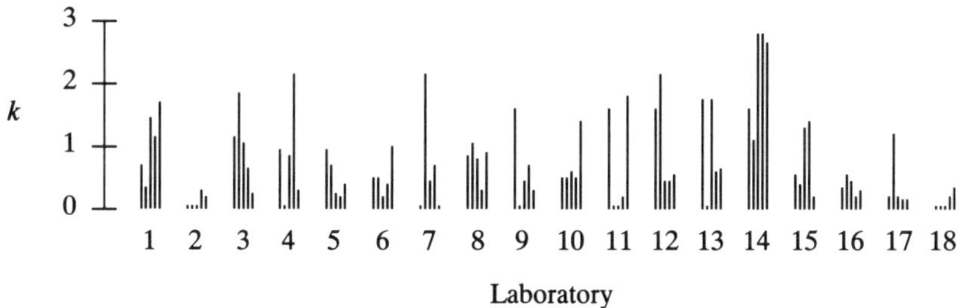

Figure 9.1. Graph of *k*-values. Data of Table 9.1.

It can be shown that the statistical distribution of k is related to that of the F statistic, according to

$$k = \sqrt{p[1 + (p - 1)/F]}\,, \tag{9.3}$$

where F is the F-ratio with $(n - 1)$ and $(n - 1)(p - 1)$ degrees of freedom.

The k statistic as applied above involves a choice between rows and columns. In our case, we did the pooling over rows. This is a rational choice whenever the rows represent samples from the same population, such as laboratories or animals of the same species. It is *not* a rational choice when they represent systematically different entities, such as serum samples with noticeably different amounts of beta-carotene (as in the case for the *columns* of Table 9.1) or as for either the rows or the columns of Table 2.1 (if there had been replication within cells in that table).

In classical analysis of variance, a "within-cell" measure of variability is calcutated as the pooled value of all s_{ij} over the entire table. Clearly, this procedure is another example of the inadvisability of using classical methods of analysis without looking in detail at the data. Let us note that analysis of variance essentially always involves pooling measures of variability to obtain the sums of squares which it compares by means of F-tests. The reader is strongly advised never to accept results without a detailed study of the data, as explained in this book.

To calculate $(s_r)_j$, we have pooled the standard deviations in column j, but our k graph has looked in close detail at this pooling for the entire table of data and, in fact, demonstrated that it was not entirely valid.

The reader may verify that in all the previous chapters as well as in the present one, we tried, as much as possible, to verify every assumption made in the course of the analysis. The reader may also note that every graph and table included the totality of the data, not just a portion of it. This is in conformance with a statement made by R.A. Fisher a number of times. For example, in his article (Fisher, 1955) he states: In one respect inductive reasoning is more strict than is deductive reasoning, since in the latter any item of the data may be ignored and valid inferences may be drawn, from the rest: i.e. from any selected subset of the set of axioms used, whereas in inductive reasoning the whole of the data must be taken into account. This seems to be very difficult to be understood by workers trained in deductive methods only, though more easily understood by statisticians. The political principle that anything can be proved by statistics arises from the practice of presenting only a selected subset of the data available.

References

Fisher, R.A. (1955). Statistical methods and scientific research. *J. Royal Statist. Soc.*, 8, 69–78.

10

Interlaboratory Testing

10.1 Introduction

There are two reasons, quite different from each other, for which interlaboratory studies are conducted. The first is a desire to monitor *laboratories* to see to what extent they can obtain the same or, at least, similar values. Such experiments are generally called "proficiency studies." Here, the laboratories themselves are our primary concern.

The second type of interlaboratory study is designed to study not so much the laboratories, as the *method of measurement*. In this type of study we still involve a number of laboratories, but we are not interested in them individually. They are just a means for studying the method.

We will discuss separately these two types of studies in this chapter.

10.2 Proficiency Studies

In a number of scientific or technological fields, for example, in clinical testing or in testing concrete, rubber, or paper, it is not uncommon for a number of laboratories (often between 50 and 150 laboratories) to participate in a proficiency test. A central agency distributes specimens of two materials to all the laboratories. Each laboratory performs replicate tests (often between two and five replicates) on each of the two materials. The results are sent to the central agency which performs a statistical analysis that will allow each laboratory to judge itself by comparing its own results to those of other laboratories. The reason for using two materials rather than one will soon become apparent.

10.3 An Example of Proficiency Testing

Table 10.1 presents the data obtained for a gloss measurement on paint chips by 30 laboratories, each of which made 4 replicate measurements on each of 2 materials (Leete, 1991). We will call the materials A and B and represent the laboratories by the numbers 1 through 30.

Table 10.1. 60 Degree Gloss of Paint Chips

Lab	A					B		
1	63.30	63.40	63.80	64.10	74.50	74.80	75.10	75.40
2	66.60	66.60	66.60	66.30	77.20	77.70	77.70	77.70
3	62.30	61.40	62.50	62.60	74.40	73.30	73.80	73.90
4	65.40	62.20	65.50	64.90	76.00	76.90	77.80	77.40
5	65.60	65.50	65.20	64.90	76.00	77.40	79.10	77.60
6	68.30	67.70	68.10	68.60	80.70	80.40	81.10	80.70
7	64.90	66.20	64.80	66.00	77.00	77.10	77.40	77.30
8	63.30	62.70	65.00	64.80	75.50	75.90	77.10	77.20
9	66.10	64.60	66.10	65.10	77.30	77.80	76.80	76.60
10	63.30	62.10	61.20	64.10	74.70	74.90	72.00	72.20
11	63.00	63.60	63.10	63.80	74.80	75.10	74.80	75.10
12	63.30	62.10	61.20	64.10	74.70	74.90	72.00	72.20
13	65.90	65.80	65.80	65.70	77.70	77.30	78.70	77.70
14	66.20	65.30	65.70	66.80	77.10	77.80	77.10	77.70
15	65.30	65.50	64.00	64.60	77.40	76.30	76.30	75.80
16	65.90	66.00	66.10	66.20	77.30	76.60	77.80	77.50
17	65.70	65.60	65.10	65.70	77.10	77.10	77.80	77.40
18	66.30	66.10	66.20	65.90	77.70	77.20	77.50	77.90
19	65.50	65.50	65.70	65.30	76.20	76.50	76.40	76.60
20	65.30	64.80	64.80	65.10	76.70	76.20	76.70	76.20
21	62.30	63.50	64.50	64.30	72.50	72.40	75.00	75.50
22	65.40	65.40	65.60	65.40	77.80	77.40	78.10	77.60
23	65.50	66.40	65.30	65.90	76.00	76.60	76.30	76.40
24	66.80	65.90	66.40	66.40	77.30	77.80	77.40	77.50
25	67.70	66.60	67.60	67.10	79.30	78.40	79.40	78.20
26	69.20	70.00	69.50	69.00	79.90	80.40	80.40	80.50
27	67.40	66.90	67.10	66.50	79.00	78.40	78.80	78.90
28	65.60	66.70	65.60	65.80	77.60	75.40	77.80	76.70
29	64.60	66.30	64.10	65.10	76.30	77.20	75.30	75.40
30	66.17	67.07	65.93	66.87	77.37	77.77	77.33	78.07

It is, of course, desirable that each of the materials be as homogeneous as possible. The reason for using two materials goes back to W.J. Youden (1959) who pointed out that a quantity that could represent within-laboratory variability can be obtained from the inclusion of two materials, even if no replicate measurements were made (i.e., if $n = 1$, in the notation of Chapter 9). Indeed, Youden was concerned that replicate measurements would not really be independent of each other, inasmuch as the laboratory would be influenced (consciously or unconsciously) by what the first measurement was in writing down a value for each subsequent measurement. However, in the present set of data, replicates were made, and we shall see how the measure of within-laboratoly variability obtained from them compares with the value obtained without the use of replicates.

Youden suggested that the two materials should be similar to each other, in the sense that

1. they have approximately (but *not* entirely) the same value for the measured property (gloss in our example),
2. they be similar in their physico-chemical matrix. This implies that any physical or chemical property of the materials, other than the one measured, be also the same or approximately the same for both materials.

It is not always practically feasible to satisfy both of these requirements, as our example shows. Indeed, the two materials are *not* close in their gloss values. Nevertheless, we will proceed with this example and present a method for analyzing them, after sketching Youden's recommended analysis.

10.3 Youden's Split-Level Analysis

We first present an outline of the analysis based on Youden's recommendations. We will, for this purpose, ignore the fact that there are replicate measurements ($n > 1$) and treat the *averages* of replicates as if they were single measurements. Table 10.2 presents, in the first three columns, the laboratory code number and the averages for A and B, which we represent by x_i and y_i.

The next two columns contain, for each laboratory, the *average* of the test results x_i and y_i, i.e., $(x_i + y_i)/2$ and the difference $y_i - x_i$. Youden's idea was that the *variability* of the differences, from laboratory to laboratory, was due to the *within-laboratory* errors.

Indeed, any systematic difference between laboratories would not affect these differences, because it would cancel out in the difference.

This can be seen by means of the equations

$$x_i = M + L_i + \delta_i, \tag{10.1}$$

$$y_i = N + L_i + \varepsilon_i, \tag{10.2}$$

where M and N are the true values for A and B, L_i is the *systematic* error of laboratory i (which is assumed to be the same for both materials), and δ_i and ε_i represent only *within-laboratory* variability.

Youden considers the deviations of x_i and y_i from their respective medians. Representing these deviations by X_i and Y_i, we have, approximately,

$$X_i = L_i + \delta_i, \tag{10.3}$$

$$Y_i = L_i + \varepsilon_i. \tag{10.4}$$

Table 10.2. Basis for Youden Analysis of Table 10.1

Lab	Averages		$\dfrac{x + y}{2}$	$y - x$
	x	y		
1	63.650	74.950	69.3000	11.3000
2	66.525	77.575	72.0500	11.0500
3	62.200	73.850	68.0250	11.6500
4	65.275	77.475	71.3750	12.2000
5	65.300	77.525	71.4125	12.2250
6	68.175	80.725	74.4500	12.5500
7	65.475	77.200	71.3375	11.7250
8	63.950	76.725	70.1875	12.4750
9	65.475	77.125	71.3000	11.6500
10	58.500	67.700	63.1000	9.2000
11	63.375	74.950	69.1625	11.5750
12	62.675	73.450	68.0625	10.7750
13	65.800	77.850	71.8250	12.0500
14	66.000	77.425	71.7125	11.4250
15	64.850	76.450	70.6500	11.6000
16	66.050	77.300	71.6750	11.2500
17	65.525	77.350	71.4375	11.8250
18	66.125	77.575	71.8500	11.4500
19	65.500	76.425	70.9625	10.9250
20	65.000	76.450	70.7250	11.4500
21	63.650	73.850	68.7500	10.2000
22	65.450	77.725	71.5875	12.2750
23	65.775	76.325	71.0500	10.5500
24	66.375	77.500	71.9375	11.1250
25	67.250	78.825	73.0375	11.5750
26	69.425	80.300	74.8625	10.8750
27	66.975	78.775	72.8750	11.8000
28	65.925	76.875	71.4000	10.9500
29	65.025	76.050	70.5375	11.0250
30	66.510	77.635	72.0725	11.1250

If we plot the X_i on the x-axis and the Y_i on the y-axis, then the variance of L_i is related to the variance of the perpendicular projection of the (X_i, Y_i) points on the main bisector of the coordinate axes, and the variance of δ_i and ε_i, considered the same for both, is obtained from the variance of the $(X_i - Y_j)$ differences.

Youden's analysis is based on a number of assumptions, such as the equality of the variances of δ_i and ε_i and the assumption that L_i represents the systematic laboratory effect for both A and B.

In the following, we present an analysis that does not depend on these assumptions.

10.4 An Alternative Analysis to Youden's Model

We write the following model:

$$x_i = E(x_i) + \delta_i, \tag{10.5}$$

$$y_i = E(y_i) + \varepsilon_i; \tag{10.6}$$

$$E(y_i) = \alpha + \beta E(x_i). \tag{10.7}$$

We also assume that even though we do not know the variances of δ_i *and* ε_i, we know the *ratio, t^2* of their variances:

$$t^2 = (\sigma_\varepsilon)^2/(\sigma_\delta)^2 \tag{10.8}$$

Because of Eqs. (10.7) and (10.8), we could apply the theory for fitting a straight line with errors in both variables (Chapter 1). Our knowledge of t^2 stems from the *replication* variances of x and y (in Youden's scheme, there is no replication) and is, therefore, really only a sample estimate. But we will ignore this and consider it a known constant.

10.5 Analysis of the Example

We first prepare an h graph and a k graph (Figs. 10.1 and 10.2). They clearly show that laboratory 10 is very low in both materials and much more variable. Unlike other applications, the present experiment is run precisely to detect and, if necessary, to *eliminate* such aberrant results. The laboratory is notified and is told that its results are not included in the calculations. The reason for this is that remedial action is expected on the part of the aberrant laboratory. We, therefore, eliminate laboratory 10 in all future calculations. Table 10.3 presents the replication standard deviations for all 29 remaining laboratories on both materials. The estimated ratio t^2 [*Eq.* (10.8)] is $(0.665/0.571)^2 = 1.36$. There is no conclusive evidence that the two pooled standard deviations are actually different, but we will continue our analysis assuming that $t^2 = 1.36$.

We have mentioned the analysis of these data based on the fitting of a straight line with errors in both variables. There is a second logical approach, to be presently discussed, which allows us to present a graphical evaluation.

We deal with the deviations of x and y from their respective mean values, $\bar{x}$ and $\bar{y}$.

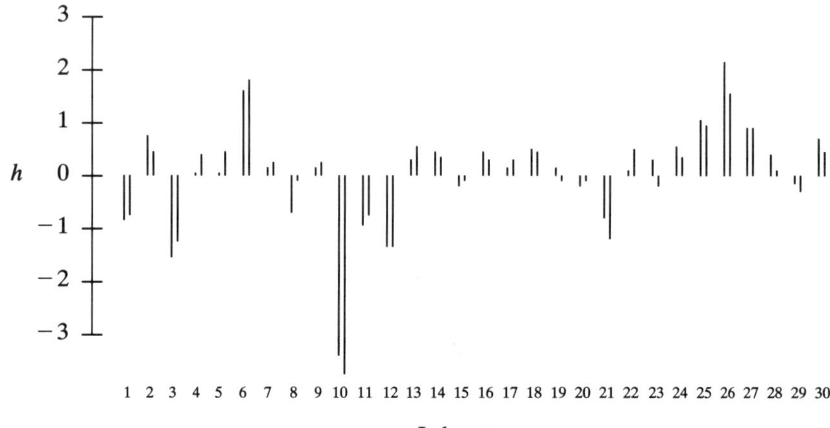

Figure 10.1. Graph of h-values. Data of Table 10.1.

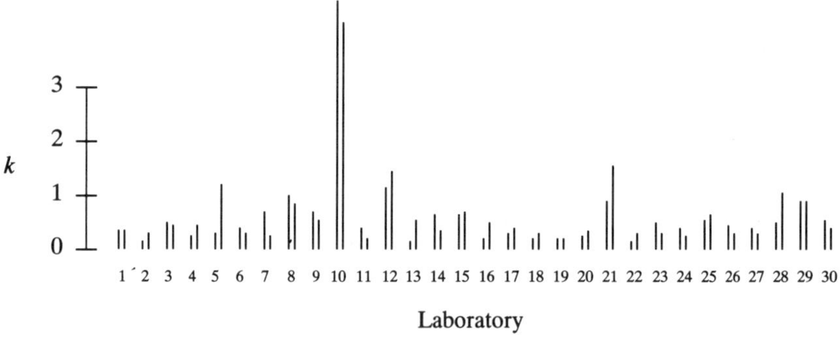

Figure 10.2. Graph of k-values. Data of Table 10.1.

Denoting these deviations by $\tilde{x}$ and $\tilde{y}$, we have

$$\tilde{x}_i = x_i - \bar{x}, \qquad \tilde{y}_i = y_i - \bar{y}. \tag{10.9}$$

Next, we introduce a variable z_i as follows:

$$z_i = \tilde{y}_i / t. \tag{10.10}$$

The error variance of z_i is $\sigma_\varepsilon^2/t^2 = t^2\sigma_\delta^2/t^2 = \sigma_\delta^2.$ Thus x_i and z_i; have equal error-variances. Under these conditions, the method of Principal Components, which is the theory underlying the Singular Value Decomposition (see Chapter 7), allows us to fit a straight line to the (x, z) points. To this effect we calculate

$$XX = \Sigma_i \tilde{x}_i^2, \tag{10.11a}$$

Table 10.3. Within-Cell Standard Deviations (Data of Table 10.1)

Lab	Standard	Deviation
1	0.3697	0.3873
2	0.1500	0.2500
3	0.5477	0.4509
4	0.2217	0.4272
5	0.3162	1.2686
6	0.3775	0.2872
7	0.7274	0.1826
8	1.1269	0.8539
9	0.7500	0.5377
10	—	—
11	0.3862	0.1732
12	1.2816	1.5631
13	0.0817	0.5972
14	0.6481	0.3775
15	0.6856	0.6758
16	0.1291	0.5099
17	0.2870	0.3317
18	0.1708	0.2986
19	0.1633	0.1708
20	0.2449	0.2887
21	0.9983	1.6299
22	0.1000	0.2986
23	0.4856	0.2500
24	0.3686	0.2160
25	0.5066	0.6131
26	0.4349	0.2708
27	0.3775	0.2630
28	0.5252	1.0935
29	0.9430	0.8888
30	0.5463	0.3515
	Pooled	
	0.571	0.665

$$XZ = \Sigma_i (\tilde{x}_i z_i), \qquad (10.11b)$$

$$ZZ = \Sigma_i z_i^2, \qquad (10.11c)$$

and we solve the following equation for λ:

$$\lambda^2 - (XX + XZ)\lambda + (XX \cdot ZZ - XZ^2) = 0.$$

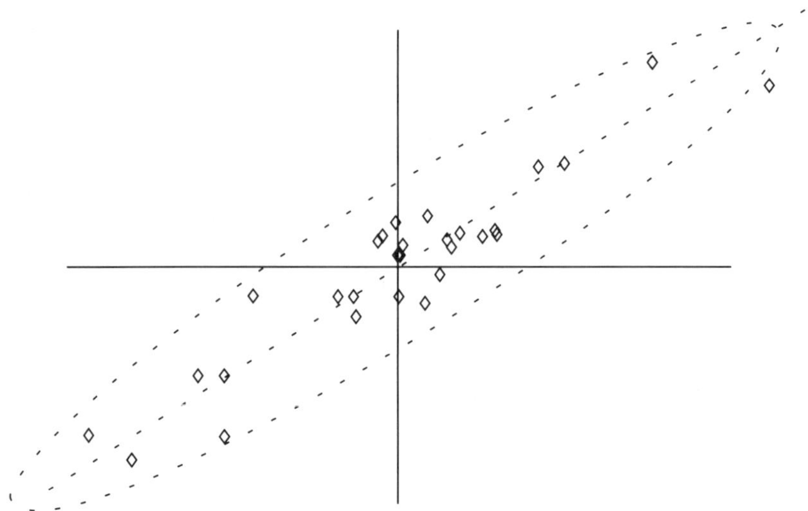

Figure 10.3. 95-Percent Ellipse for data of Table 10.1.

This yields the values, λ_1 and λ_2, where λ_1 is the larger root. We next calculate the quantity

$$\tan a = (\lambda_1 - XX)/XZ. \tag{10.12}$$

Finally, we calculate the quantity $t \tan \alpha$.

The angle whose tangent is $(t \tan \alpha)$ is the angle that the best fitting line to the (x, y) points makes with the x-axis. Thus, the tangent of that angle, say $\tan \theta$, is the desired slope of the best fitting line to the (x, y) points. We have

$$\tan \theta = t \tan \alpha. \tag{10.13}$$

It is also possible to construct an ellipse such that a desired percentage of (x, y) points will lie inside of the ellipse. This is done as follows:

Calculate a and b by the formulas

$$a = \sqrt{\frac{2\lambda_1 F}{p-2}}, \qquad b = \sqrt{\frac{2\lambda_2 F}{p-2}}, \tag{10.14}$$

where F is the critical value of the F-statistic with 2 and $(p\text{-}2)$ degrees of freedom for the desired confidence coefficient (the fraction of points desired inside of the ellipse, e.g., 95%), and p is the number of laboratories.
Also calculate the quantities

$$A = b^2\cos^2\alpha + a^2\sin^2\alpha, \tag{10.14a}$$

Table 10.4. Principal Component Analysis of Data of Table 10.1

$$s_r(\delta) = 0.5707 \qquad s_r(\varepsilon) = 0.6650 \qquad t^2 = (s_\varepsilon/s_\delta)^2 = 1.36$$

$$XX = 65.5504 \qquad ZZ = 57.5094 \qquad XZ = 57.6833$$

$$\lambda^2 - 123.0598\,\lambda + 442.4053 = 0$$

$$\lambda_1 = 119.3532 \qquad \lambda_2 = 3.7067$$

$$\tan\alpha = \frac{119.3532 - 65.5504}{57.6833} = 0.9327$$

$$\tan\theta = t\tan\alpha = \sqrt{1.36}\,(0.9327) = 1.0877$$

$$\theta = \arctan(1.0877) = 47.40° = 0.8274 \text{ radians}$$

$$a = \sqrt{\frac{2(119.3532)(3.35)}{29-2}} = 5.4422 \qquad b = \sqrt{\frac{2(3.7067)(3.35)}{29-2}} = 0.9591$$

$$A = 14.2706 \qquad B = -24.5484 \qquad C = 11.9608 \qquad D = 27.2443$$

Equation of 95% ellipse: $Ax^2 + Bxy + Cy^2 = D$

$$B = [\,2(b^2 - a^2)\sin\alpha\cos\alpha\,]/t, \qquad (10.14\text{b})$$

$$C = [\,b^2\sin^2\alpha + a^2\cos^2\alpha\,]/t^2, \qquad (10{,}14\text{c})$$

$$D = a^2 b^2. \qquad (10.14\text{d})$$

The desired ellipse satisfies the equation

$$Ax^2 + Bxy + Cy^2 = D. \qquad (10.15)$$

These calculations are illustrated in Table 10.4 for the data of Table 10.1, and a graph of the points and of the ellipse is shown in Fig. 10.3.

Each participating laboratory receives a copy of the statistical analysis as well as a copy of Fig. 10.3. By locating its own results (x, y point) on the graph, it knows exactly how it compares to all the other laboratories and whether remedial action is necessary in order to improve its analytical procedure. The exercise is repeated several times a year (usually four times), using a different pair of samples each time.

It is of interest to compare the errors in both variables, obtained as deviations from the regression line, with those provided by replication. We find the following:

Errors obtained as deviations from regression line (based on the average of four measurements):

for x: standard deviation $= 0.3705$

for y: standard deviation $= 0.4321$

Errors obtained from replicates:
    for x: standard deviation $= 0.571$
    for y: standard deviation $= 0.665$

The latter, in order to be compared with the former, should be divided by $\sqrt{4}$. Hence,
    for x: standard deviation $= 0.285$
    for y: standard deviation $= 0.332$

Thus, the deviations from the fitted line are larger than what would be expected from replication. This is not unusual and indicates that an additional source of error is present in the scatter of points about the fitted straight line when compared to the variability among replicates. This additional source amounts to about 30% of the replication error.

There is probably no better procedure for monitoring the performance of laboratories than that given by the process we have just described.

10.6 The Study of a Test Method

We have already pointed out that there are two reasons for conducting interlaboratory studies: to monitor laboratories or to study the method of measurement. We have dealt with the former in the preceding sections and now turn to the second type.

Standardization organizations, such as the American Society for Testing Materials (ASTM), the International Standardization Organization (ISO), or the Association of Official Analytical Chemists (AOAC) require that the standardization of a test method include a statement on the precision and accuracy of the method.

The *precision* of a method is a measure for the closeness of results obtained by repeating the entire measurement process on a different portion of the same material. It depends on the *mode of repetition*. Thus, repeated measurements made in a laboratory on the same day, by the same operator, and using the same instrumental setup are generally in better agreement with each other than measurements made on different days or by different operators; those, in turn, tend to be closer to each other than measurements made in different laboratories. Because there are so many different possible situations, it has been decided in general by standardization organizations to consider only two extreme situations: within the same laboratory and between laboratories. The former is termed *repeatability* and the latter *reproducibility*.

Repeatability is defined more precisely by considering repeated measurements made as much as possible by the same operator, on the same measuring equipment, as close as possible to each other in time.

10.7 The Determination of Beta-Carotene in Serum

We return to the data in Table 9.1. In the last chapter we looked at these data only from the wiewpoint of within-laboratory variability (i.e., replication error), using the k statistic. Now we study the data in a more complete manner.

We first prepare h and k graphs. These are shown in Figs. 10.4 and 10.5. The first of these exhibits the very definite systematic errors of the various laboratories. In particular, laboratories 12 and 18 are very low when compared to other laboratories, and laboratory 14 is high. These laboratories should have been investigated. In fact, we will soon see evidence that, despite their extreme positions, they followed the general pattern of the other laboratories. Rejection of such laboratories without investigation is, as we have already pointed out, an irrational procedure that accomplishes nothing positive but gives the committee a false idea of higher precision than the data show. Figure 10.5, the k-graph, also shows some anomalies that are worthy of further investigation.

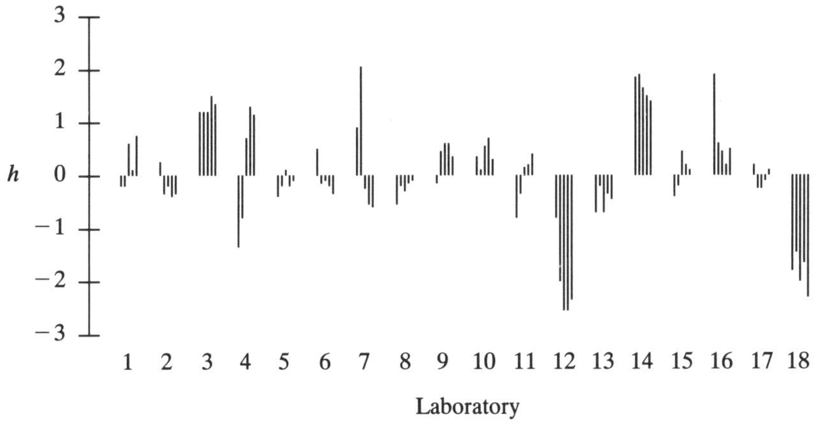

Figure 10.4. Graph of h-values. Data of Table 9.1.

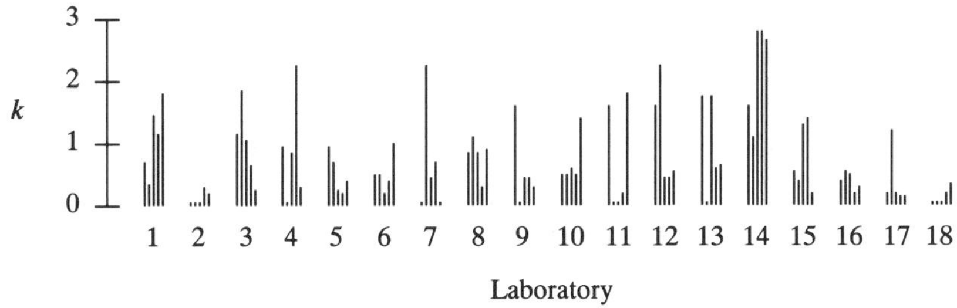

Figure 10.5. Graph of k-values. Data of Table 9.1.

Laboratory 14, which was high on the h graph, is also high (more variable) on the k graph.

In Chapter 9 we explained how to calculate the *repeatability standard deviation, s_r*. It is customary, in the analysis of interlaboratory data to calculate a second quantity called the *reproducibility standard deviation, s_R*. It is obtained by

$$s_R = \sqrt{s_r^2 + \left(s_L^2 - \frac{s_r^2}{n}\right)}, \tag{10.16}$$

where s_L is the standard deviation between the cell averages in a particular column, over laboratories; n is the number of replicates in each cell. The quantities s_r, s_L, and s_R all refer to a particular column; to simplify the notation we omitted the subscript j from these symbols. The subtraction of s_r^2/n from s_L^2 is due the fact that the variance between cell averages contains a contribution, s_r^2/n, due to within-cell variability. Because of statistical sampling errors, the quantity $s_L^2 - s_r^2/n$ may occasionally be negative. In that case, it is replaced by zero.

Figure 10.6 is a plot of s_r and s_R against the level of the material. We see very smooth relationships, despite the inclusion of the discrepant laboratories

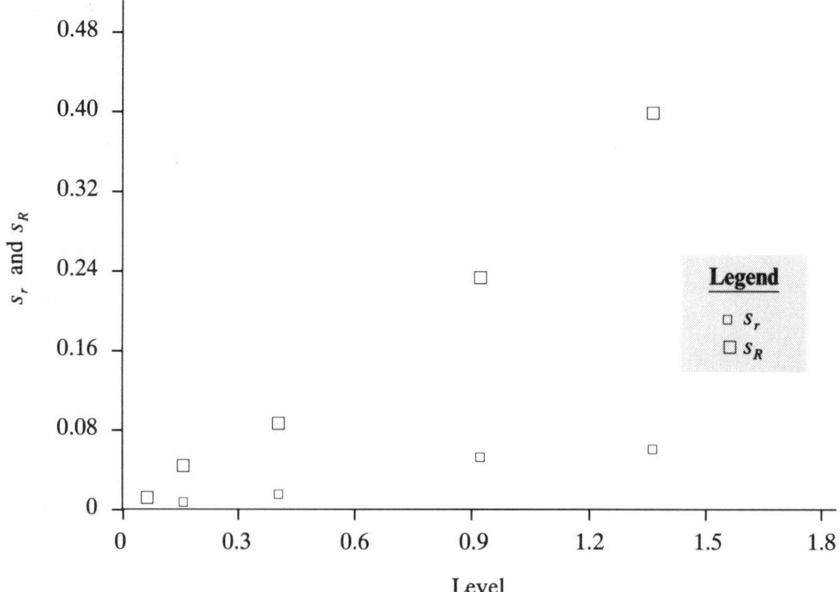

Figure 10.6. Graph of $s_r(j)$ and $s_R(j)$. Data of Table 9.1.

12, 14, and 18. We also note the fact that s_R is appreciably larger than s_r. A test that exhibits such a situation can probably be made more reproducible among laboratories by further research aimed at finding the cause for the appreciable lab-to-lab variability. Such research might also clarify the situation with regard to discrepant laboratories.

Although the preceding is generally a satisfactory evaluation of the test method (Fig. 10.4, 10.5, and 10.6), it is of interest to study more closely the structure of these data.

Figure 10.7 presents the diagnostic plot provided by the additive-multiplicative analysis (see Section 7.7). The plot indicates that the data are very nearly both row-linear and column-linear (i.e., concurrent). The analysis of variance of the residuals is shown in Table 10.5.

The preferred structure is the row-linear one, because the mean square of the residuals is the smallest. We test for concurrence by means of the F-test:

$$F_{16,51} = \frac{0.002534}{0.001149} = 2.206.$$

The F-value is significant at the 5% level and leads one to the conclusion that the concurrent model is not entirely acceptable. Nevertheless, the data

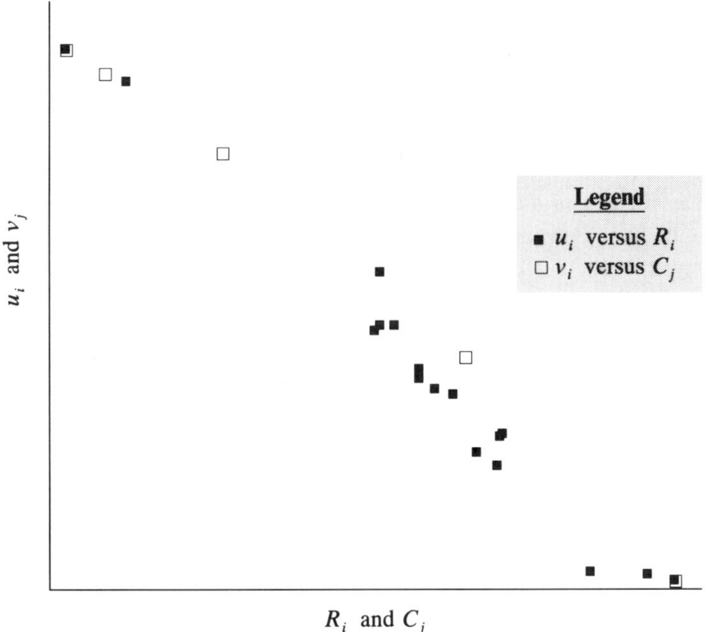

Figure 10.7. Diagnostic plot of Table 9.1.

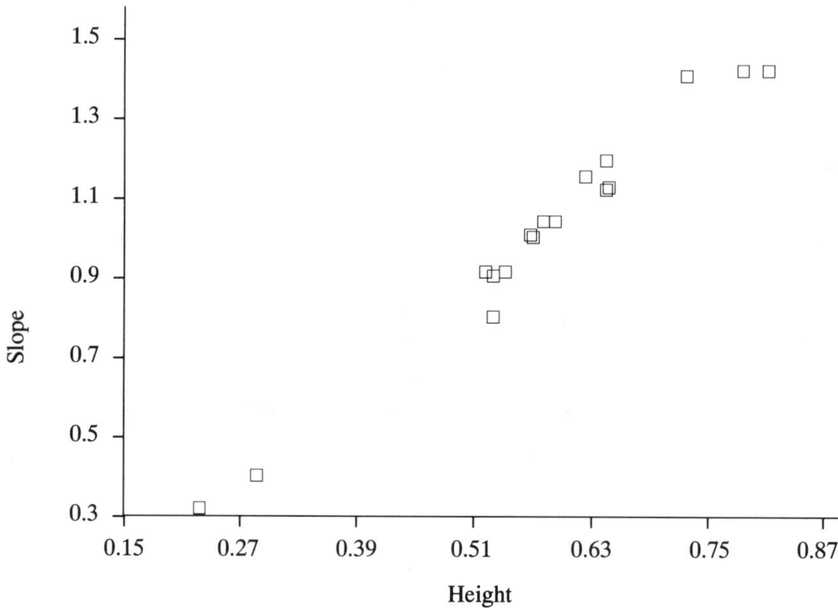

Figure 10.8. Slope versus height. Data of Table 9.1.

indicate a strong tendency for the lines representing the laboratories to meet inside a small area, albeit not in a single point. Figure 10.8 is a plot of the slopes of the lines versus the heights (ordinates corresponding to an abscissa of $\bar{x}$, the average of all x_j-values). Although these points do not lie exactly on a straight line, they are nevertheless close to a straight line whose slope is different from zero. This confirms the conclusions drawn from Fig. 10.7. It also confirms our previous remark that laboratories 12 and 18, although discrepant, nevertheless follow the structural pattern of the other laboratories. Indeed, these two laboratories are represented by the two low points on the left of Fig. 10.7:

Table 10.5. Analysis of Variance for Table 9.1

Structure	Sum of Squares of Residuals	Degrees of Freedom	Mean Square
Row-linear	0.05858	51	0.001149
Col-linear	0.08940	64	0.001397
Concurrent	0.09913	67	0.001480
Additive	1.78273	68	0.026220

They are very close to the straight line drawn through the remaining points. What this indicates is that these two laboratories did something systematically different from the others; they did *not* make sloppy measurements.

References

Leete, C.G. (1991). Personal communication.

Youden, W.J. (1959) Graphical Diagnosis of Interlaboratory Tests. Summary Technical Repart No. 2312. U.S. Dept. of Commerce. National Bureau of standards.

11

Missing Values

11.1 Introduction

Occasionally, it happens that a two-way table of measurements is incomplete in the sense that one or more cells are missing. This occurs when one or more measurements are defective and have to be discarded, or when the measurements in question simply were not made. Such an occurrence makes the techniques of analysis we have discussed, as well as other techniques found in a variety of textbooks, inapplicable. It is, therefore, not surprising that formulas have been developed for "estimating" a missing cell (see, e.g., Snedecor and Cochran, 1972). The reader is advised *not* to use this formula, unless he is certain that his data follow an additive structure. Indeed, the formula postulates such a structure and is erroneous if the structure is not additive.

11.2 Theoretical Considerations

How a missing value is to be estimated depends on the structure of the data. If the data are row-linear or column-linear, it is relatively easy to estimate the missing value. We illustrate the procedure for a row-linear structure.

Let y_{ij} be the general element in a $p \times q$ row-linear table.

Suppose that the element y_{mn} (i.e., the element in the mth row and the nth column) is missing.

1. Calculate all column averages, *omitting* row m in the averaging process. Call these column averages x'_j (for the jth column).
2. Regress linearly the elements in row m (except the missing y_{mn}) on the corresponding x'_j (omitting, of course, x'_n). Let a and b be the estimated intercept and slope respectively of this line.
3. Then an estimate for the missing element is

$$y' = a + bx'_n. \tag{11.1}$$

If more than one cell is missing, then the x_j' are calculated using only rows that are complete (that have no missing values).

11.3 A Numerical Example

Table 11.1 presents data obtained by Wood and Martin (1964) on the compressibility of natural rubber. We coded the data as indicated in the table. Measurements were made on peroxide-cured rubber and on unvulcanized rubber at several temperatures and pressures. No measurement is available for the unvulcanized rubber at 20°C and 500 kg/cm^2 pressure. We have already used some of the data (the peroxide-cured rubber only) previously (Table 2.1). We now use the complete set, which includes the missing cell.

Table 11.1. Specific Volume of Rubber

Temp. (°C)	Pressure (kg/cm^2)[a]					
	500	400	300	200	100	0
	Peroxide Cured					
0	137	178	219	263	307	357
10	197	239	282	328	376	427
20	256	301	346	394	444	498
25	286	330	377	426	477	532
	Unvulcanized					
0	54	93	136	179	225	272
20	–	218	264	314	364	417
25	202	248	295	345	396	451

[a] Tabulated value = (spec. vol. -1.05) $\times 10^4$.

Even though the data were obtained on two different samples, it is possible to treat the data as a row-linear 7×6 table, after estimating the missing value.

In Table 11.2, we indicate the averages x_j', for all six columns, where x_j' was calculated on the complete rows (i.e., leaving out the row that has the missing cell). The regression of this row (omitting the first element) on the corresponding x_j' yields the intercept a and slope b:

$$a = -23.0553,$$

$$b = 1.0424.$$

Using x_1', we have

$$y' = a + bx_1' = 23.0553 + (1.0424)(188.667) = 173.6.$$

Table 11.2. Column Averages of Table 11.1 Omitting Row with Missing Value

Column	A	B	C	D	E	F
x_j'	188.67	231.50	275.83	322.50	370.83	422.83

Table 11.3. Parameters for Row-Linear Fit (Data of Table 11.1 with Calculated Value)

Row	Height	Slope	Standard Deviation of Slope
1	243.50	0.9305	0.0033
2	308.17	0.9767	0.0017
3	373.17	1.0253	0.0022
4	404.67	1.0454	0.0023
5	159.83	0.9291	0.0039
6	291.77	1.0361	0.0029
7	322.83	1.0568	0.0012

Table 11.4. Analysis of Variance of Residuals for Two Fits (Data of Table 11.1 with Calculated Value)

Structure	Sum of Squares of Residuals	Degrees of Freedom	Mean Square
Row-linear	7.640503	24	0.3184
Two Column-linear Fits	$\left\{ \begin{array}{l} 1.6293\,[a] \\ 0.8801\,[b] \end{array} \right.$	$\left. \begin{array}{l} 10 \\ 5 \end{array} \right\}$	0.1673

[a] Peroxide cured.
[b] Unvulcanized.

If we had used the missing value formula of Snedecor and Cochran, we would have obtained 179.4, which, as can readily be verified, is not a reasonable value.

Further analysis of the data, using the row-linear model, is now possible.

Figure 11.1 is a diagnostic plot of u_i versus row average and v_j versus column average (see Chapter 7). Different symbols have been used for the two types of rubber. We see that for each rubber *separately*, the row markers are linear. The column markers for all the data are also linear. The indications are that we have two concurrent sets, one for each rubber.

The h graph is shown in Fig. 11.2 and the h' graph in Fig. 11.3. The h graph shows two distinct sets, corresponding to the two rubbers (rows 1 to 4, and rows 5 to 7). The h' graph shows that each row is very probably linearly related to the column averages over *all* rows. Thus, the row-linear model is applicable to the entire set despite the presence of the two rubbers.

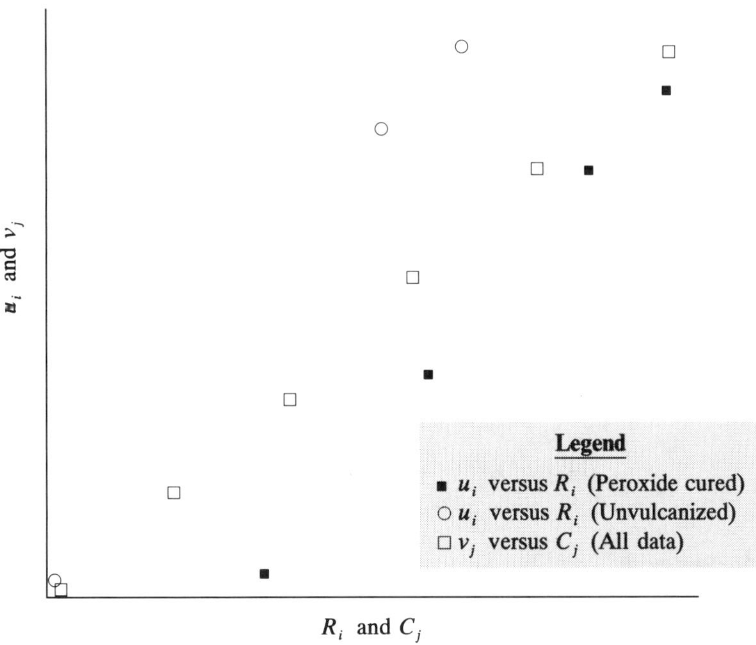

Figure 11.1. Diagnostic plot of Table 11.1 (with estimated value for missing plot).

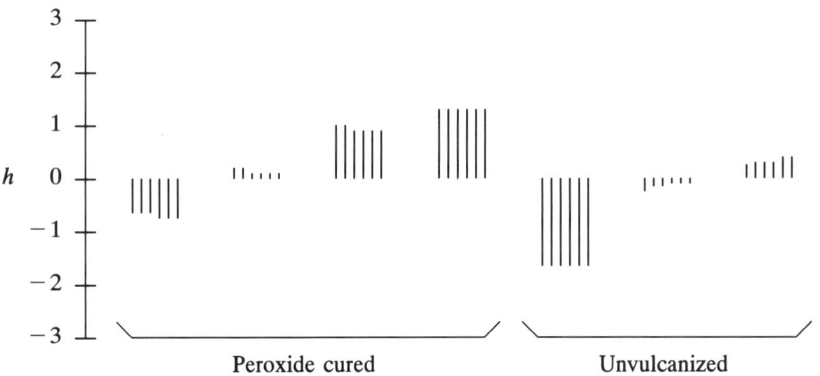

Figure 11.2. Graph of h-values. Data of Table 11.1.

Table 11.3 shows the averages and slopes (versus the column averages) for all rows. We can compare the residuals based on row-linear fits. The results are shown in Table 11.4. Clearly, the column-linearity is stronger than the row-linearity. The preferable model for these data is that of two column-linear structures, one for each rubber.

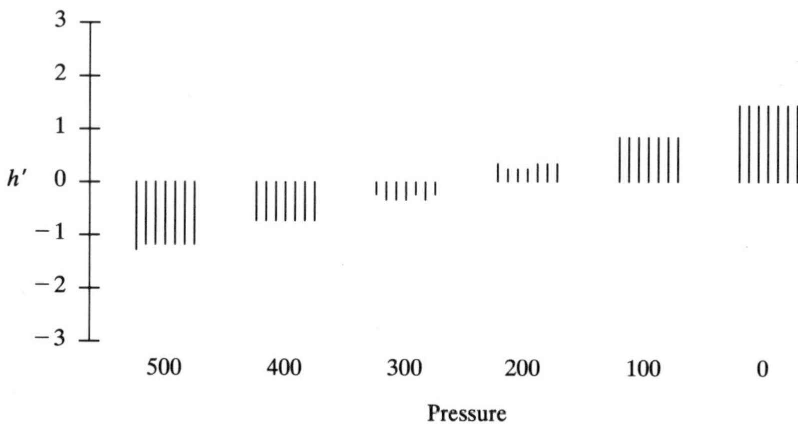

Figure 11.3. Graph of h'-values. Data of Table 11.1.

References

Snedecor, G.W. and W.G. Cochran (1972). *Statistical Methods*. The Iowa State University Press, Ames.

Wood, L.A. and G.M. Martin (1964). Compressibility of natural rubber at pressures below 500 kg/cm^2. *J. Res. Nat. Bur. Stand., Sec. A. Phys. Chem.*, 68A 259–268.

12

Curve Fitting

12.1 Introduction

With the exception of the first chapter, this book, so far, has been concerned with two way arrays of measurements. In the present chapter, we depart, temporarily, from two way tables because in some of the subsequent chapters we will be concerned with fitting structural parameters (such as row and column means) obtained from an *internal* analysis of the data, to the corresponding labels.

Such fitting often involves *curves* rather than straight lines. We, therefore, describe a method for fitting curves. It is customary to accomplish the fitting of a curve by means of polynomial expressions such as

$$y = a_0 + a_1 x + a_2 x^2 + a_3 x^3 + \cdots + a_k x^k . \tag{12.1}$$

Indeed, *any* set of points can be fitted by such an equation.

If $k + 1 =$ number of points, then the fit is *exact,* i.e., the fitted curve (polynomial) goes through every point. In most cases, this is totally undesirable because the points are generally affected by errors of measurement. Therefore, the curve should be smooth and in the vicinity of each of the points, but not necessarily go through every point.

The quadratic function $(k = 2)$ has the equation

$$y = A + Bx + Cx^2 \tag{12.2}$$

and is, of course, much more flexible than the straight line.

The quadratic function would be a reasonable candidate for fitting many curves, but it has a serious drawback. Indeed, any pair of points with abscissas equal to $-B/(2C) - \Delta$ and $-B/(2C) + \Delta$ have the same ordinate y, regardless of the value of Δ. The value $x = -B/(2C)$ is such that points at equal distance from it, to its left and to its right, have the same ordinate. This symmetry is oflen a hindrance to making the quadratic a close fit to a set of experimental points. A modification of quadratic can be found to make it more flexible.

12.2 The Box–Cox Transformation

Box and Cox (1964) proposed a function of the type

$$x_t = (x^\alpha - 1)/\alpha \qquad (12.3)$$

for purposes with which we need not be concerned. The symbol x_t stands for "x transformed." The function contains a parameter, α. It is to be noted that as α approaches zero, x_t approaches the natural logarithm of x. Thus, the function $\ln(x)$ is a special case of Eq. (12.3) when α is very small.

We propose to replace x in Eq. (12.2) by x_t, with an appropriate choice of α, and then use the equation

$$y = A + B[(x^\alpha - 1)/\alpha] + C[(x^\alpha - 1)/\alpha]^2 \qquad (12.4)$$

to fit a set of (x, y) points. Our problem is to find the best value of α in each case.

Finding an appropriate value of α [for Eq. (12.3)] and fitting a set of (x, y) points by Eq. (12.4) will be referred to as fitting a QFP, which stands for Quadratic Four Parameter fit. The four parameters are α, A, B, and C.

We propose that before applying QFP, we standardize both x and y. Although this step is not always essential, it is very useful and sometimes even necessary. We recommend that it always be used. The exact procedure, in its entirety, is discussed below.

12.3 Procedure for Fitting Quadratic Four Parameter Curve

The following procedure has been found to give excellent results in many cases.

1. Standardize both x_j and y_j into u_j and v_j as follows:

$$u_j = 2 + (x_j - \bar{x})/\sqrt{XX}, \qquad (12.5)$$

$$v_j = (y_j - \bar{y})/\sqrt{YY}, \qquad (12.6)$$

where $\bar{x}$ and $\bar{y}$ are the averages of the x_j and the y_j, respectively, and

$$XX = \Sigma_j (x_j - \bar{x})^2, \qquad YY = \Sigma_j (y_j - \bar{y})^2. \qquad (12.7)$$

The reason for adding 2 to $(x_j - \bar{x})/\sqrt{XX}$ is that the quantity would be between -1 and $+1$ without it, and the Box–Cox transformation cannot be made on a negative quantity.

2. A number of α-values are tried, say 20 values between -5 and $+5$. For each of these, the Box–Cox transformation is made:

$$z_j = (u_j^\alpha - 1)/\alpha \qquad (12.8)$$

and v_j is fitted to z_j by means of the quadratic equation

$$\hat{v}_j = A + Bz_j + Cz_j^2, \qquad (12.9)$$

where $\hat{v}_j$ is an estimate for v_j. For details of the fitting of this quadratic, see the Appendix. On the computer, this trial-and-error procedure is quite feasible and fairly rapid.

For each α, the correlation coefficient between the observed v_j and the fitted $\hat{v}_j$ is calculated. Denote the α for which the correlation is highest by α_0.

3. A new set of, say, 20 α-values is tried for α between $\alpha_0 - 1$ and $\alpha_0 + 1$. Again, correlation coefficients are calculated as above, and the α-value with the highest correlation is chosen as the final α. Equations (12.8) and (12.9) then provide the desired values for A, B and C.

4. By means of Eq. (12.6), v_i is transformed back into a y_i value. Thus, a y-value is found for each x_i-value. Moreover, Eqs. (12.5) and (12.8) allow us to express Eq. (12.9) as a functional relation between x and y.

12.4 First Numerical Example

We return to the set exhibited in Table 1.1, which deals with the specific volume of rubber at various temperatures. In Chapter 1 we saw that a straight line fit was not satisfactory because a slight degree of curvature was indicated. Table 12.1 exhibits the data, coded as indicated. Applying the procedure of

Table 12.1. Data of Table 1.1 with Coded y-Values

Temperature	Specific Volume (Coded)[a]
10	19.652
15	16.724
20	13.692
25	10.528
30	7.259
35	3.874
40	0.390

[a] Tabulated values = (specific volume -0.94) $\times 10^3$.

Table 12.2. QFP Fit to Data of Table 12.1

$$\alpha = 1.2$$

$$U(J) = 2 + (X(J) - 25)/26.45752$$

$$V(J) = [Y(J) - 10.30271]/17.00102$$

$$Z(j) = [U(j)^{\alpha} - 1]/\alpha$$

$$\widehat{V}(J) = A + BZ(J) + CZ(J)^2$$

$$\widehat{Y}(J) = 10.30271 + 17.00102\,\widehat{V}(j)$$

$$A = 0.9167103, \quad B = -0.8004946 \quad C = -3.246058E-02$$

$X(J)$	$U(J)$	$Z(J)$	$\widehat{V}(J)$	$Y(J)$	$\widehat{Y}(J)$
10.00000	1.43305	0.44998	0.54993	19.65200	19.65207
15.00000	1.62204	0.65565	0.37791	16.72400	16.72754
20.00000	1.81102	0.86619	0.19898	13.69200	13.68556
25.00000	2.00000	1.08116	0.01330	10.52800	10.52885
30.00000	2.18898	1.30025	-0.17901	7.25900	7.25934
35.00000	2.37796	1.52315	-0.37787	3.87400	3.87846
40.00000	2.56695	1.74963	-0.58323	0.39000	0.38719

Section 12.3 to the data, we obtain the results shown in Table 12.2. The fit is excellent, far better than the straight line fit.

It may be observed that we needed four parameters to fit seven points. However, our points represent density for temperature values between 10° and 40°C. The equation we have fitted can be used to estimate density values for *any* temperature between 10° and 40°C. It is, therefore, much more than a fit to seven points. For example, if we wish to know the density at 32°C, we obtain density (32°C) = 0.94592 and this value is entirely reliable because Eq. (12.4) with our calculated parameters is smooth and monotonic in the entire range of temperature values (10° to 40°C) (see Mandel, 1981).

A polynomial of degree 6 would have fitted the seven data points perfectly but would be useless as a prediction equation for other points. In any curve-fitting procedure, it is imperative to also consider the smoothness and, if necessary, the monotonicity of the fit, not only its success in reproducing the experimental points.

12.5 A Second Example

We present a second example because in some cases the application of QFP requires a prior transformation. Table 12.3 represents the critical values of Student's-*t* statistic, at 5% level of significance, for 20 values of the degrees of freedom, DF.

Table 12.3. Critical Values of Student's t (5 % Level)

DF	t	DF	t
2	4.3027	14	2.1448
3	3.1825	16	2.1199
4	2.7764	18	2.1009
5	2.5706	20	2.0860
6	2.4469	25	2.0595
7	2.3646	30	2.0423
8	2.3060	40	2.0211
9	2.2622	60	2.0003
10	2.2281	120	1.9799
12	2.1788	∞	1.9600

The difficulty of fitting this set is that we have one point (the last one) at $DF = \infty$. Obviously, we cannot plot all the 20 DF-values on an axis. The difficulty can be solved by considering as the x variable, not DF, but 1/DF. If $x = 1/DF$, then the range of x-values is 0 to 0.5. If we apply QFP to the x-values, with the observed y-values (Student's t), we obtain the results shown in Table 12.4. Here again, we have fitted far more than 20 values. Our fit is smooth and monotonic and covers *any* DF value between 2 and infinity. For example, for $DF = 23$, our formula yields a t-value of 2.069; the tabular value (not used in our fit) is 2.0687. For $DF = 160$, our formula yields a t-value of 1.977, a very acceptable value.

A legitimate question concerns the choice of the transformation $x_t = 1/x$. Why did we pick this particular transformation? The answer is that other transformations could have been chosen and would have led to entirely satisfactory results. For example, $x_t = 1/x^{1.2}$ could have been chosen. Table 12.5 shows the results with this choice. The main requirement is to reduce the infinite range of the original DF variable to a finite one. The fact that different transformations can be chosen attests to the robustness of the QFP procedure when carried out in accordance with Section 12.3.

12.6 Conclusion

The QFP fit is quite versatile, as we have shown above, in fitting a variety of (x, y) relationships. In subsequent chapters we will apply it to the fitting of structural parameters of two-way tables to the corresponding labels, in cases where labels are present.

Chap. 12. Curve Fitting

Table 12.4. QFP Analysis of Table 12.3 [a]

$$\alpha = 3$$

$$U(J) = 2 + [X(J) - 0.1187565]/0.5412286$$

$$V(J) = [Y(J) - 2.356675]/2.392595$$

$$Z(j) = [U(j)^{\alpha} - 1]/\alpha$$

$$\widehat{V}(J) = A + BZ(J) + CZ(J)^2$$

$$\widehat{Y}(J) = 2.356675 + 2.392595\,\widehat{V}(j)$$

$$A = -0.3973223, \quad B = 0.1359822, \quad C = 9.159042E-03$$

x	x Transf	U	Z	$\widehat{V}$	Y	$\widehat{Y}$
2.000	0.500	2.704	6.260	0.813	4.303	4.301
3.000	0.333	2.396	4.254	0.347	3.183	3.187
4.000	0.250	2.242	3.426	0.176	2.776	2.778
5.000	0.200	2.150	2.980	0.089	2.571	2.570
6.000	0.167	2.089	2.703	0.037	2.447	2.446
7.000	0.143	2.045	2.515	0.003	2.365	2.363
8.000	0.125	2.012	2.380	-0.022	2.306	2.304
9.000	0.111	1.986	2.277	-0.040	2.262	2.261
10.000	0.100	1.965	2.197	-0.054	2.228	2.227
12.000	0.083	1.935	2.080	-0.075	2.179	2.178
14.000	0.071	1.913	1.999	-0.089	2.145	2.144
16.000	0.063	1.896	1.939	-0.099	2.120	2.119
18.000	0.056	1.883	1.893	-0.107	2.101	2.100
20.000	0.050	1.873	1.857	-0.113	2.086	2.086
25.000	0.040	1.854	1.793	-0.124	2.060	2.060
30.000	0.033	1.842	1.751	-0.131	2.042	2.043
40.000	0.025	1.827	1.699	-0.140	2.021	2.022
60.000	0.017	1.811	1.648	-0.148	2.000	2.002
120.000	0.008	1.796	1.598	-0.157	1.980	1.982
∞	0.000	1.781	1.548	-0.165	1.960	1.962

[a] With trasformation $x = 1/DF$.

References

Box, G.E.P. and D.R. Cox (1964). An analysis of transformations. *J. Roy. Statist. Soc., Ser. B* 26, 211–252.

Mandel, J. (1981). Fitting curves and surfaces with monotonic and non-monotonic four parameter equations. *J. Res. Nat. Bur. Stand.*, 86, 1–25.

Table 12.5. QFP Analysis of Table 12.3 [a]

$$\alpha = 2.1$$

$$U(J) = 2 + [X(J) - 8.577351E - 02]/0.4668719$$

$$V(J) = [Y(J) - 2.356675]/2.392595$$

$$Z(j) = [U(j)^\alpha - 1]/\alpha$$

$$\widehat{V}(J) = A + BZ(J) + CZ(J)^2$$

$$\widehat{Y}(J) = 2.356675 + 2.392595\,\widehat{V}(j)$$

$$A = -0.5477253, \quad B = 0.2977104, \quad C = 2.568621E\ 02$$

x	x Transf	U	Z	$\widehat{V}$	Y	$\widehat{Y}$
2.000	0.435	2.749	3.504	0.811	4.303	4.297
3.000	0.268	2.389	2.490	0.353	3.183	3.201
4.000	0.189	2.222	2.071	0.179	2.776	2.785
5.000	0.145	2.127	1.846	0.090	2.571	2.571
6.000	0.116	2.066	1.709	0.036	2.447	2.443
7.000	0.097	2.024	1.616	0.001	2.365	2.358
8.000	0.082	1.993	1.550	−0.025	2.306	2.298
9.000	0.072	1.970	1.501	−0.043	2.262	2.254
10.000	0.063	1.951	1.463	−0.057	2.228	2.219
12.000	0.051	1.925	1.408	−0.078	2.179	2.171
14.000	0.042	1.907	1.370	−0.092	2.145	2.137
16.000	0.036	1.893	1.343	−0.102	2.120	2.114
18.000	0.031	1.883	1.323	−0.109	2.101	2.096
20.000	0.027	1.875	1.307	−0.115	2.086	2.082
25.000	0.021	1.861	1.279	−0.125	2.060	2.058
30.000	0.017	1.852	1.262	−0.131	2.042	2.043
40.000	0.012	1.842	1.241	−0.139	2.021	2.025
60.000	0.007	1.832	1.222	−0.146	2.000	2.008
120.000	0.003	1.823	1.205	−0.152	1.980	1.993
∞	0.000	1.816	1.191	−0.157	1.960	1.982

[a] With trasformation $x = 1/DF^{1.2}$.

Appendix: Fitting a Quadratic Function

Given a set of N (z, v) values, we are to find values of A, B, and C in the equation

$$v_i = A + Bz_i + Cz_i^2 \qquad [\text{see Eq. (2.8)}].$$

One procedure is to use orthogonal polynomials, as follows. Write

$$v = \alpha P_0 + \beta P_1 + \gamma P_2, \tag{12.10}$$

where

$$P_0 = 1,$$
$$P_1(i) = z_i - \bar{z}, \tag{12.11}$$
$$P_2(i) = [P_1(i)]^2 - aP_1(i) - b.$$

Here, $\bar{z}$ is the average of the z-values. It is seen that P_0, P_1, and P_2 are of degree 0, 1, and 2 respectively in z.

If SS_2 and SS_3 are defined as

$$SS_2 = \Sigma_i (z_i - \bar{z})^2, \tag{12.12}$$
$$SS_3 = \Sigma_i (z_i - \bar{z})^3, \tag{12.13}$$

then

$$a = SS_3/SS_2, \qquad b = SS_2/N. \tag{12.14}$$

It can be shown that estimates of α, β, and γ are

$$\hat{\alpha} = \bar{v}, \tag{12.15}$$

where $\bar{v}$ is the average of the v-values,

$$\hat{\beta} = \frac{\Sigma_i [v_i P_1(i)]}{\Sigma_i [P_1(i)]^2}, \tag{12.16}$$

and

$$\hat{\gamma} = \frac{\Sigma_i [v_i P_2(i)]}{\Sigma_i [P_2(i)]^2}. \tag{12.17}$$

13

Two-Way Tables with Quantitative Labels

13.1 Introduction

We are now in a position to fit a two-way table with quantitative labels. As we mentioned earlier, we propose to perform such a fit in two steps. First, we make an analysis of the data *inside* the table, ignoring the labels. Second, we fit the structural parameters provided by the analysis to the labels by means of curve-fitting procedures. We will see that for the examples we will present, QFP is a satisfactory procedure for the curve-fitting phase.

First, we make an important observation. Any fit to the data we can produce will be *empirical*. It would, indeed, be presumptuous to pretend that we have come up with a theoretically valid fit. There is no need to apologize for this because, for example, most of the thermodynamic models for equations of state are more empirical than theoretical; and this is also the case for much of science.

Suppose that we are confronted with a large body of data, or even a smaller body of data that covers a large range of values of the independent variables. Then, if we succeed in representing these data by empirical equations that reproduce the data very closely and, allow us to interpolate for values of the independent variables that are *not* present in the data, then we consider the effort highly successful.

Now, in the preceding chapters we have dealt in detail with specific structures, especially row- and column-linear structures. From the view point of empirical fitting, we do not need to identify the data with any of these structures.

Suppose that, in accordance with Chapter 7, we have successfully represented a set of data by the equations

$$\hat{y}_{ij} = -M + R_i + C_j + \theta \cdot u_i \cdot v_j , \qquad (13.1a)$$

$$y_{ij} = \hat{y}_{ij} + d_{ij} , \qquad (13.1b)$$

where $\hat{y}_{ij}$ is an estimate for the observed y_{ij}, M, R_i, and C_j are estimators for the additive parameters, and $\theta \cdot u_i \cdot v_j$ is the first term obtained by applying the SVD procedure to the residuals from the additive fit. Then, if the residuals d_{ij} are sufficiently small, Eq. (13.1a) is an acceptable empirical fit for the data. The shortcoming of Eq. (13.1a) is that it expresses $\hat{y}_{ij}$ as a function of "structural" parameters, such as R_i, C_j, u_i, and v_j instead of expressing it in terms of *labels*.

In this chapter we turn to the task of expressing R_i and u_i in terms of the row labels, and C_j and v_j as functions of the column labels. We will also suppose that there are reasonably smooth relations between the row label and R_i, the row label and u_i, the column label and C_j, and the column label and v_j.

13.2 The Density of Aqueous Solutions of Alcohol

The density of aqueous solutions of ethanol was determined with great precision by Osborne, Mc Kelvy, and Bearce (1913) as a function of concentration of alcohol and of temperature. The coded data are shown in Table 13.1.

Table 13.1. Density of Aqueous Solutions of Ethanol[a]

Conc.	Temperature (°C)						
	10	15	20	25	30	35	40
30.086	139.652	136.724	133.692	130.528	127.259	123.874	120.390
39.988	122.415	118.851	115.219	111.502	107.727	103.876	99.946
49.961	101.704	97.847	93.922	89.938	85.880	81.784	77.588
59.976	79.323	75.289	71.202	67.049	62.842	58.570	54.233
70.012	55.989	51.848	47.640	43.378	39.060	34.683	30.240
80.036	31.882	27.642	23.363	19.030	14.646	10.202	5.694

[a] Coded: Tabulated values = (density -0.82) $\times$ 10^3.

Fitting an additive-multiplicative structure to the data (see Chapter 7), we obtain the graph displayed in Fig. 13.1. The relations, although not linear, are nonetheless very smooth. The parameters are displayed in Table 13.2.

Applying the QFP procedure (see Chapter 12) to R_i and u_i as functions of the row label and to C_j and v_j as functions of the column label, we obtain the parameters displayed in Table 13.3. The individual points and the fitted curves are shown in Fig. 13.2. We recall the equations for the QFP fit: y_i is fitted to a quantity z_i, which is a known function of x_i, as follows:

$$u_i = 2 + \frac{x_i - \bar{x}}{s_x}, \qquad (13.2)$$

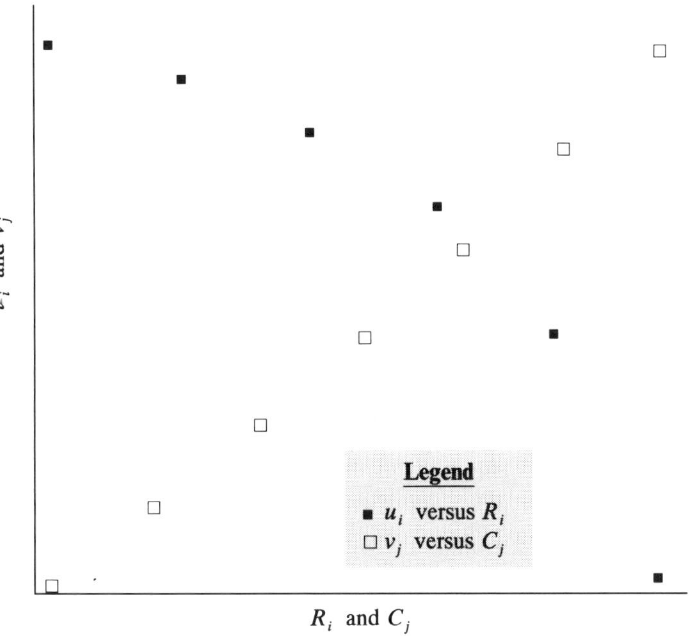

Figure 13.1. Diagnostic plot of data of Table 13.1.

$$z_i = (u_i^\alpha - 1)/\alpha, \tag{13.3}$$

$$\hat{y}_i = A + B_i z_i + C z_i^2. \tag{13.4}$$

For our example, the value of M is 76.76483 and the value of θ is 5.109576. Thus, Eq. (13.1a) becomes

$$\hat{y}_{ij} = -76.76483 + R_i + C_j + 5.109576 \, (u_i \cdot v_j). \tag{13.5}$$

Table 13.2. Structural Parameters for Fit of Table 13.1
$$(\hat{Y}_{ij} = -76.76483 + R_i + C_j + 5.109576(u_i \cdot v_j))$$

	Rows			Columns	
Label	R_i	u_i	Label	C_j	v_j
30.086	130.3027	−0.784751	10	88.49418	0.590284
39.988	111.3623	−0.231493	15	84.70016	0.377437
49.961	89.80901	0.052136	20	80.83968	0.173148
59.976	66.92973	0.220473	25	76.90418	−0.018318
70.012	43.26258	0.334702	30	72.90234	−0.201870
80.036	18.92272	0.408932	35	68.83151	−0.376020
			40	64.68185	−0.544654

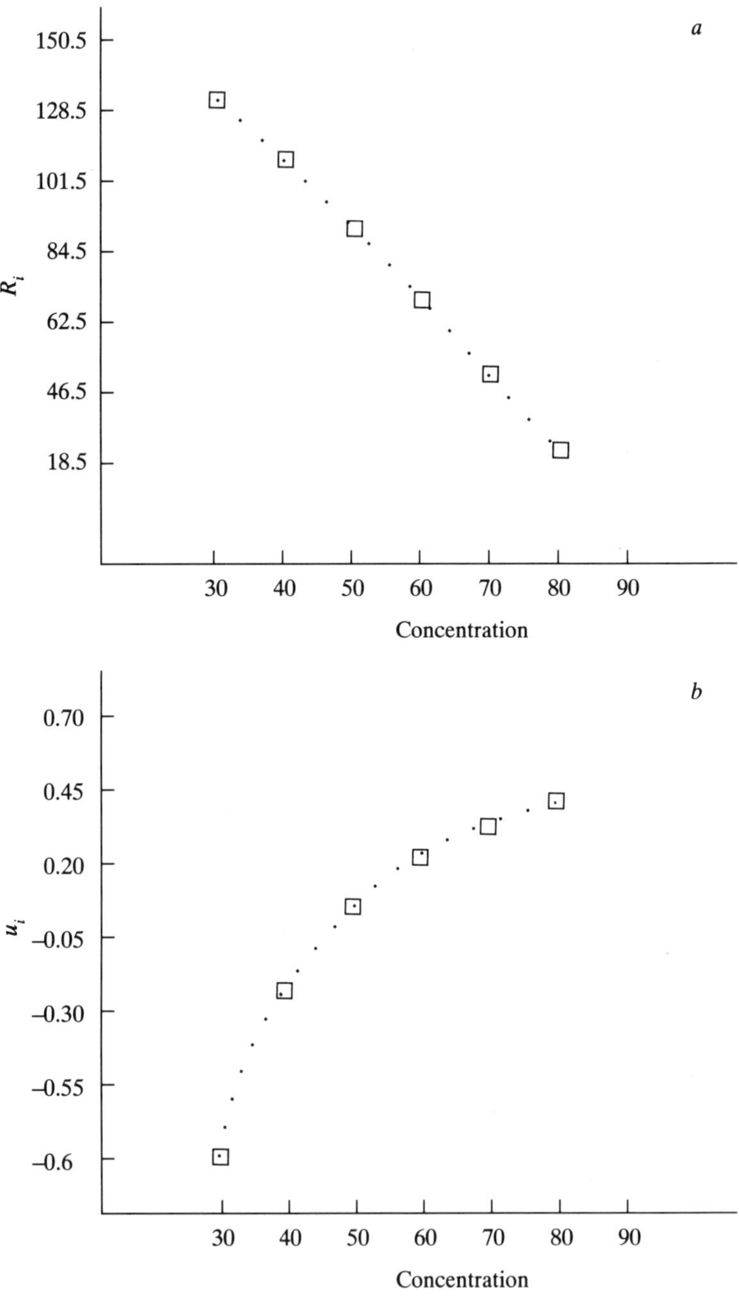

Figure 13.2. QFP fits for R_i, u_i, C_j, v_j. Data of Table 13.1.

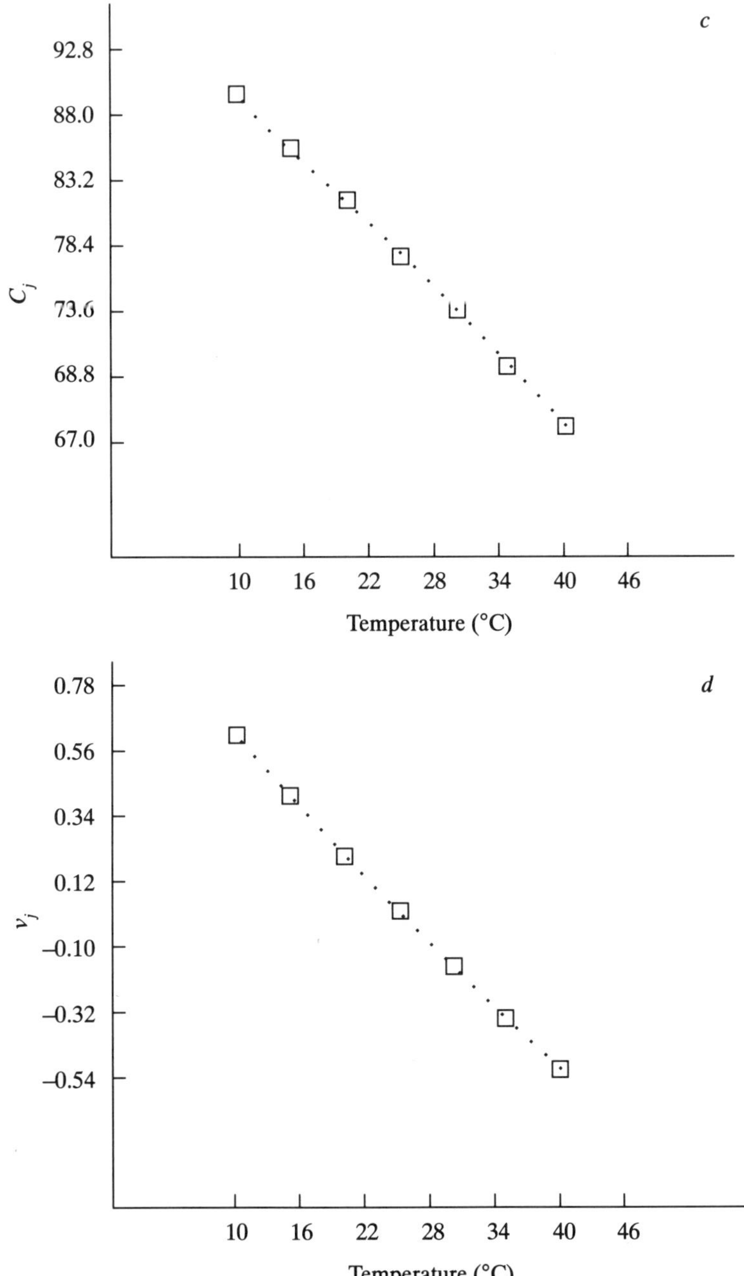

Figure 13.2. *cont.* QFP fits for R_i, u_i, C_j, v_j. Data of Table 13.1.

Table 13.3. QFPs for Fit of R_i, u_i, C_j, and v_j (see Table 13.2)

Quantity	Rows		Columns	
	R	u	C	v
$\bar{x}$	55.00984	55.00984	25.0	25.0
s_x	41.81365	41.81365	26.45752	26.45752
α	2.0	-2.4	0.8	0.1
A	157.5812	-3.892524	96.93704	1.137207
B	-57.61679	16.5216	-19.12747	-1.368568
C	3.274226	-13.46793	-2.693784	-0.3363773

The parameters $\bar{x}$, s_x, α, A, B, and C in Table 13.3 are calculated in accordance with Eqs. (13.2)–(13.4) where x is either the row label or the column label and $\hat{y}$, is the valve fitted by these equations for and R_i, u_i, C_j or v_j. The residuals of the fit $(y_{ij} - \hat{y}_{ij})$ are shown in Table 13.4, where y now stands for any entry in Table 13.1, and $\hat{y}$ for its estimate. The range of values in Table 13.1 is $139.652 - 5.694 = 133.958$, and the largest residual in Table 13.4 is 0.22, which is 0.16% of the range. In terms of percent relative deviation from the original (uncoded) value, we obtain the values shown in Table 13.5. Interpolation will be possible, and successful, for any concentration value between $10°$ and $40°$C.

Table 13.4. Residuals for Fitted Model[a]

-0.092	-0.085	-0.065	-0.066	-0.065	-0.077	-0.088
0.220	0.190	0.176	0.159	0.166	0.177	0.189
-0.085	-0.090	-0.096	-0.094	-0.098	-0.071	-0.075
-0.115	-0.110	-0.101	-0.100	-0.093	-0.089	-0.087
0.100	0.117	0.117	0.115	0.113	0.109	0.097
-0.025	-0.027	-0.024	-0.025	-0.026	-0.032	-0.046

[a] In same units as Table 13.1.

Table 13.5. Percent Deviation for Fitted Model[a]

-0.010	-0.009	-0.007	-0.007	-0.007	-0.008	-0.009
0.023	0.020	0.019	0.017	0.018	0.019	0.021
-0.009	-0.010	-0.011	-0.010	-0.011	-0.008	-0.008
-0.013	-0.012	-0.011	-0.011	-0.010	-0.010	-0.010
0.011	0.013	0.013	0.013	0.013	0.013	0.011
-0.003	-0.003	-0.003	-0.003	-0.003	-0.004	-0.006

[a] With [(Measured–Fitted)/Measured], where Measured and Fitted are in original density units. For example, for cell $(1,1)$, the residual (Measured − Fitted) = -0.000092, which is -0.010% of the original density 0.959652.

13.3 Conclusions

For data of high precision, such as those of Table 13.1, our two-step procedure for obtaining an empirical fit to the data, as well as for interpolation at all values of the independent variables in the ranges of the experiment, is highly satisfactory. Occasionally, a fit based on theory may also yield good results, but such instances are rare. We emphasize again that our procedure should not be condemned because of its need for a fairly large number of parameters (see Table 13.3). This is more than compensated for by the ability to interpolate successfully anywhere in the original table of values. It is worth noting that our numerical example involves a relatively small number of values: a total of 42 measurements.

The procedure we have outlined can be used successfully for fitting equations of state.

Reference

Osborne, N.S., E.C. McKelvy, and H.W. Bearce, (1913). Density and Thermal expansion of Ethyl alcohol and its mixtures with water. *Bull. Bur. Stand.,* 9 328–468.

14

Applications to Real Data Sets

14.1 Introduction

In this chapter, we demonstrate the applicability of our analysis to a variety of data found in the literature. Some of these will have quantitative labels, others will not. We will treat such examples accordingly.

Table 14.1. Total Solids in Wet Brewer's Yeast[a]. Drying Time (hours)

Sample	3	6	9
1	3.24	3.16	2.96
	3.56	3.26	3.01
2	3.92	3.81	3.76
	3.86	3.80	3.75
3	9.13	8.86	8.70
	9.28	8.79	8.75
4	8.35	8.11	7.94
	8.29	8.24	7.99
5	5.51	5.06	4.84
	5.53	5.11	4.80
6	6.63	6.61	6.60
	6.65	6.57	6.55
7	9.29	8.96	8.84
	9.28	9.12	9.03
8	8.76	8.39	8.23
	8.72	8.43	8.27
9	8.03	7.86	7.72
	8.11	7.84	7.79
10	6.61	6.32	6.21
	6.77	6.23	6.13

[a] (Coded: Tabulated value = total solids −10.

14.2 Total Solids in Wet Brewers' Yeast

Table 14.1 presents the data (Bennett and Franklin, 1954). Figure 14.1 is the AMA plot (see Chapter 7). It shows linearity in the column markers and, therefore, a row-linear structure. Table 14.2 exhibits the analysis of variance of residuals which confirms the superiority of the row-linear structure over other structures. Table 14.3 exhibits the parameters for the row-linear fit.

Table 14.2. Analysis of Variance of Residuals (Data of Table 14.1)

Structure	Sum of Freedom of Residuals	Degrees of Freedom	Mean Square
Row-linear	0.01639	9	0.00182
Column-linear	0.1740	16	0.01088
Concurrent	0.1766	17	0.01039
Additive	0.1784	18	0.00991

Figures 14.2 and 14.3 show the h and k plots, by groups of samples. The variability of samples is strongly indicated in the h graph. The k plot shows that some samples (especially 1 and 7) exhibit more variability than others.

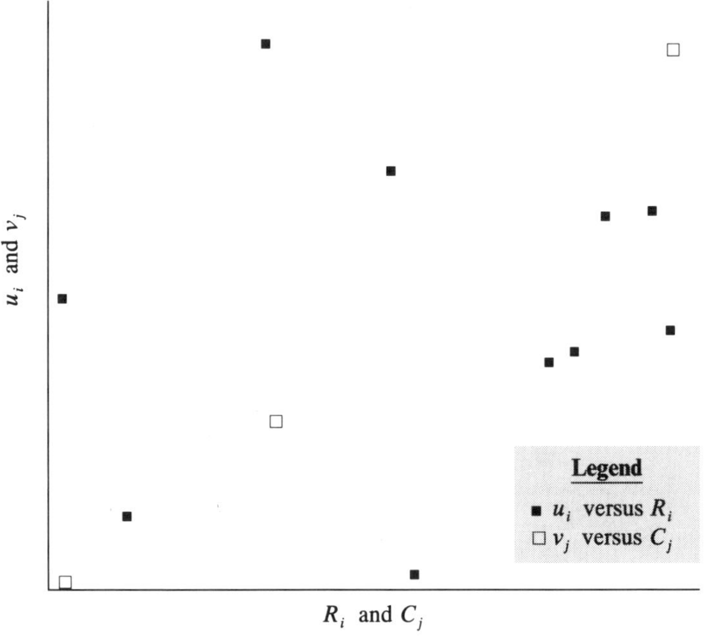

Figure 14.1. Diagnostic plot of data of Table 14.1.

Table 14.3. Parameters of Row-Linear Fit (Data of Table 14.1)

Sample	Height	Slope
1	3.1983	1.0434
2	3.8167	0.3514
3	8.9183	1.2894
4	8.1533	0.8836
5	5.1417	1.8189
6	6.6017	0.1738
7	9.0867	0.9235
8	8.4667	1.2865
9	7.8917	0.8311
10	6.3783	1.3985

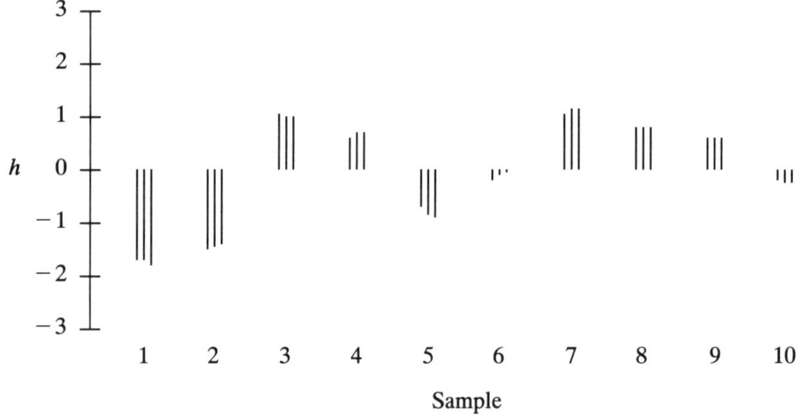

Figure 14.2. Graph of h-values. Data of Table 14.1.

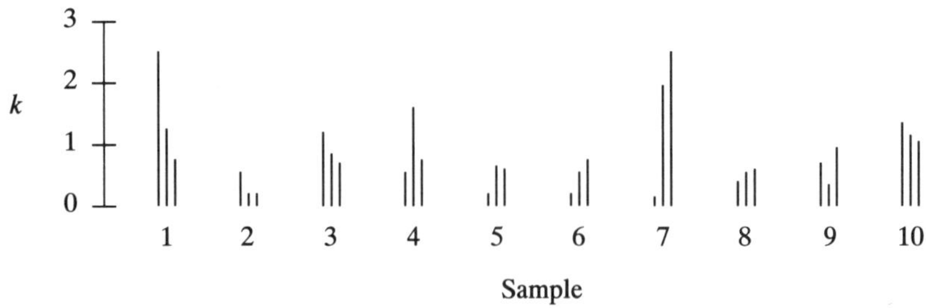

Figure 14.3. Graph of k-values. Data of Table 14.1.

An h' plot, shown in Fig. 14.4, shows sharp differences caused by changes in drying time. The effect of this label, which is quantitative, is shown in Fig. 14.5 where a quadratic has been fitted to the data.[1] The equation of this fit is given by

$$\text{Column average} = a + b(\text{DT}) + c(\text{DT})^2, \qquad (14.1)$$

where DT is the drying time and

$$a = 7.342003, \qquad b = -0.141418, \qquad c = 0.006472. \qquad (14.2)$$

The residuals of the row-linear fit are exhibited in Table 14.4.

Table 14.4. Residuals from Row-Linear Fit (Data of Table 14.1)

-0.018	0.052	-0.034
-0.001	0.002	-0.001
0.015	-0.043	0.028
-0.019	0.056	-0.036
-0.005	0.014	-0.009
0.002	-0.005	0.003
0.004	-0.011	0.007
0.002	-0.007	0.004
0.003	-0.009	0.006
0.017	-0.049	0.032

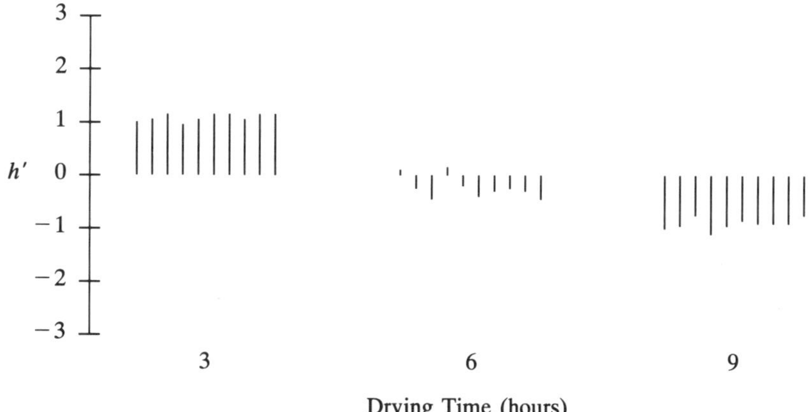

Figure 14.4. Graph of h'-values. Data of Table 14.1.

[1] With only three points, it is impossible to fit a QFP.

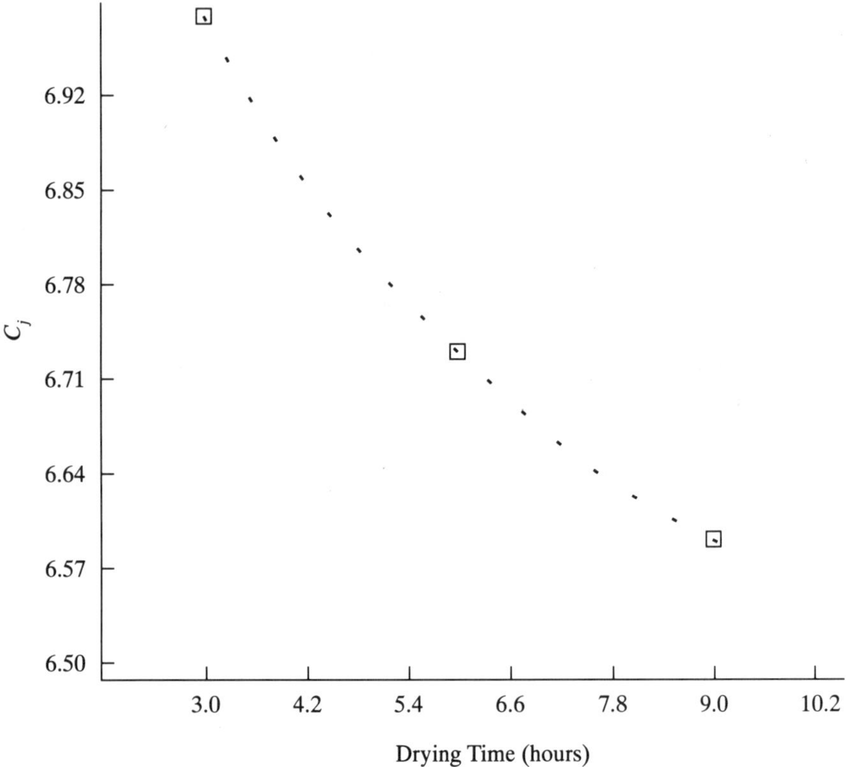

Figure 14.5. Quadratic fit of C_j versus drying time. Data of Table 14.1.

The items worth noting are that the data are definitely neither additive nor concurrent. Turkey's one degree of freedom test for nonadditivity would have provided an erroneous model for the structure of these data. It is also worth noting that the data have provided an answer to the question of the effect of drying time. However, because only three times were considered, the quadratic fit to drying time is not entirely convincing.

14.3 Absorption of Gamma Rays by Aluminum

An experiment was performed to determine the absorption of gamma rays by aluminum. Plates of aluminum were placed between a radioactive source and a Geiger–Muller counter. The distance between source and counter was evaluated in a rough manner. Unlike the experiment described in Section 14.2, this set of data was obtained in a classroom experiment and is not of high precision. The thickness of the absorbing aluminum was determined by the number of plates of aluminum used. Table 14.5 list the data, in units of $\log_{10}$ (pulses per second) for six distances (rows) and six thicknesses of interposed aluminum. The experiment was repeated a second time. Thus, our data have 12 rows and 6 columns. A systematic difference can be observed between the two experiments.

Table 14.5. Absorption of Gamma Rays by Aluminum[a]

	Distance	Number of Aluminum Plates					
		0	1	3	6	7	10
First	3.8	1.849	1.834	1.818	1.811	1.790	1.777
	5.3	1.666	1.632	1.613	1.600	1.603	1.597
	6.1	1.532	1.509	1.482	1.476	1.454	1.447
experiment	9.1	1.257	1.249	1.224	1.204	1.211	1.179
	12.4	1.017	0.986	0.971	0.966	0.960	0.943
	15.9	0.830	0.793	0.801	0.755	0.754	0.763
Second	3.8	1.946	1.916	1.913	1.884	1.887	1.871
	5.3	1.754	1.732	1.723	1.698	1.696	1.674
	6.1	1.662	1.632	1.624	1.592	1.588	1.579
experiment	9.1	1.382	1.344	1.341	1.312	1.311	1.290
	12.4	1.142	1.118	1.118	1.106	1.066	1.066
	15.9	0.954	0.924	0.913	0.906	0.892	0.870

[a] Coded: Tabulated value $= \log_{10}$ (pulses/sec). Corrected for background noise.

Plots of h and h' are shouwn in Figs. 14.6 and 14.7. An analysis of variance of residuals for various structures is shown in Table 14.6. It is apparent that an additive model is acceptable. This is confirmed by a diagnostic plot of row and column markers (not shown here). Consequently, for these data, we fit the model

$$\hat{y}_{ij} = -M + R_i + C_j . \tag{14.3}$$

Table 14.7 lists the row and column labels and R_i and C_j, as well as M. Table 14.8 lists the QFP parameters for R_i and C_j. Table 14.9 lists the residuals obtained by fitting Eq. (14.3) with the estimates for R_i and C_j obtained from Table 14.8. They indicate that our fit is quite acceptable.

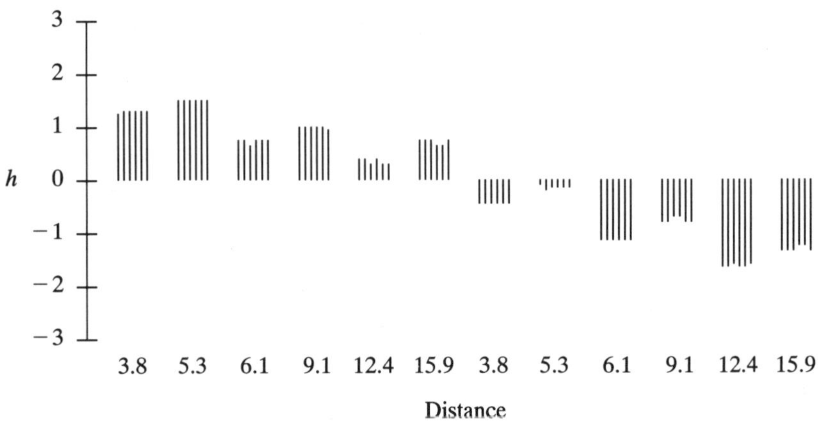

Figure 14.6. Graph of h-values. Data of Table 14.5.

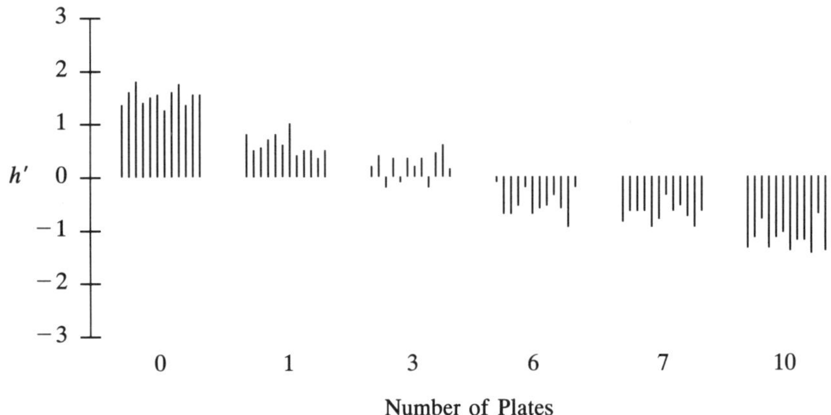

Figure 14.7. Graph of h'-values. Data of Table 14.5.

Table 14.6. Analysis of Variance of Residuals (Data of Table 14.5)

Structure	Sum of Squares[a]	Degrees of Freedom	Mean Square[a]
Row-linear	2.661	44	0.06049
Column-linear	2.975	50	0.05951
Concurrent	3.028	54	0.05607
Additive	3.032	55	0.05513

[a] Multiplied by 1000.

Table 14.7. Structural Parameters for Data of Table 14.5

Rows		Columns	
Row Label	R_i	Column Label	C_j
3.8	1.8132	0	1.4159
	1.9028		
5.3	1.6185	1	1.3991
	1.7128		
6.1	1.4833	3	1.3784
	1.6128		
9.1	1.2207	6	1.3592
	1.3300		
12.4	0.9738	7	1.3527
	1.1060		
15.9	0.7827	10	1.3380
	0.9098		
$M = 1.3722$			

Table 14.8. QFP Parameters for R_i and C_j (Data of Table 14.5)

Value	R_i	C_j
$\bar{x}$	8.7667	4.5
s_x	14.6965	8.5732
α	-1.4	-1.4
A	5.0337	1.5302
B	-10.1906	-0.4476
C	4.0001	0.1700

Table 14.9. Residuals from QFP Fitting of Parameters (Data of Table 14.5)

-0.048	-0.048	-0.044	-0.031	-0.047	-0.049
0.049	0 034	0.051	0.042	0.050	0.045
-0.029	-0.049	-0.047	-0.040	-0.033	-0.028
0.059	0.051	0.063	0.058	0.060	0.049
0.069	-0.078	-0.084	-0.070	-0.087	-0.083
0.061	0.045	0.058	0.046	0.047	0.049
-0.054	-0.048	-0.052	-0.052	-0.040	-0.061
0.071	0.047	0.065	0.056	0.060	0.050
-0.058	-0.075	-0.069	-0.054	-0.055	-0.061
0.067	0.057	0.078	0.086	0.071	0.062
-0.058	-0.081	-0.052	-0.078	-0.074	-0.054
0.066	0.050	0.060	0.073	0.064	0.053

14.4 Two Further Examples

We now discuss two examples taken from the same paper (Waxler, Weir, and Schampf, 1964). The reason for discussing two examples in the same section will soon be apparent.

The first example is shown in Table 14.10. The data are refractive index values of carbon tetrachloride at various pressures and wavelengths at a temperature of 24.80°C. Figure 14.8 is a diagnostic plot of u_i versus R_i and of v_j versus C_j (see Chapter 7).

Table 14.10. Refractive Index of Carbon Tetrachloride at 24.80°C[a]

Pressure (bars)	Wavelength (Å)[b]							
1	0.476	0.556	0.791	1.270	1.322	1.400	1.511	1.626
276.9	1.762	1.846	2.089	2.583	2.635	2.721	2.831	2.951
523.6	2.706	2.790	3.037	3.542	3.598	3.674	3.794	3.917
797.4	3.599	3.683	3.934	4.404	4.509	4.591	4.712	4.835
1116.7	4.502	4.591	4.847	5.371	5.429	5.515	5.639	5.767

[a] Coded: Tabulated value = (refractive index $- 1.45) \times 100$.
[b] See "column labels" in Table 14.13.

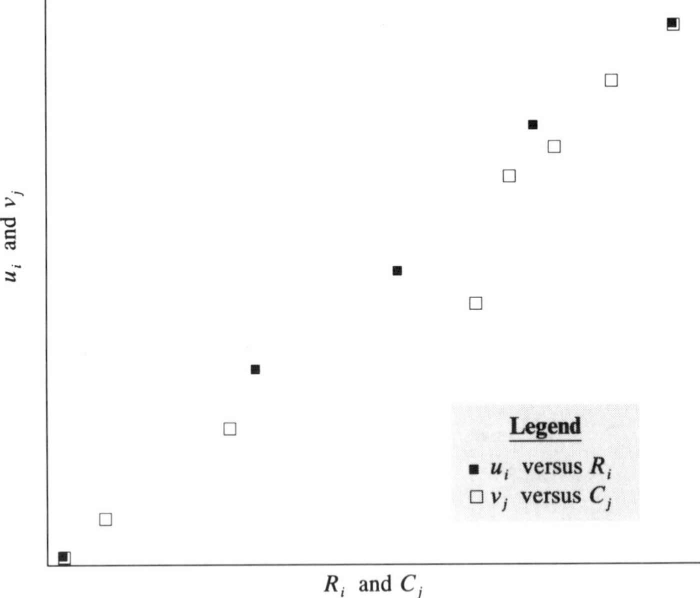

Figure 14.8. Diagnostic plot. Data of Table 14.10.

It is apparent that one column marker (for the fourth column) is completely out of line. Figure 14.9 is a plot of the residuals from an additive fit by columns. Clearly, the value corresponding to row 4 and column 4 is far lower than might be expected. This points to the value 4.404 in Table 14.10. If this value is eliminated and considered as a missing value, and the technique of Chapter 11 for missing values is applied, we obtain the estimate 4.450. With the value 4.450, the diagnostic plot, shown in Fig. 14.10, is well behaved. It is very likely that the value 4.404 is a copying error for a measurement that might have been 4.440. At any rate, we forego any further discussion of this data set and turn to Table 14.11, which is a similar set, obtained by the same authors, but at a temperature of 34.5°C.

The diagnostic plot is shown in Fig. 14.11 and points to approximate (but not complete) row- and column-linearity. Because we merely wish to obtain an empirical fit, we adopt the more general model

$$y_{ij} = -M + R_i + C_j + \theta \cdot u_i \cdot v_j + d_{ij}. \tag{14.4}$$

Table 14.11. Refractive Index of Carbon Tetrachloride at 34.50°C[a]

Pressure (bars)	Wavelength (Å)[b]							
1	0.890	0.965	1.197	1.675	1.723	1.802	1.908	2.020
255.4	2.164	2.243	2.481	2.975	3.022	3.104	3.214	3.329
480.9	3.102	3.182	3.426	3.928	3.980	4.059	4.172	4.292
756.1	4.045	4.130	4.381	4.896	4.949	5.031	5.146	5.271
1119.9	5.098	5.186	5.441	5.973	6.028	6.109	6.229	6.360

[a] Coded: Tabulated value = (refractive index − 1.44) × 100.
[b] See "column labels" in Table 14.13.

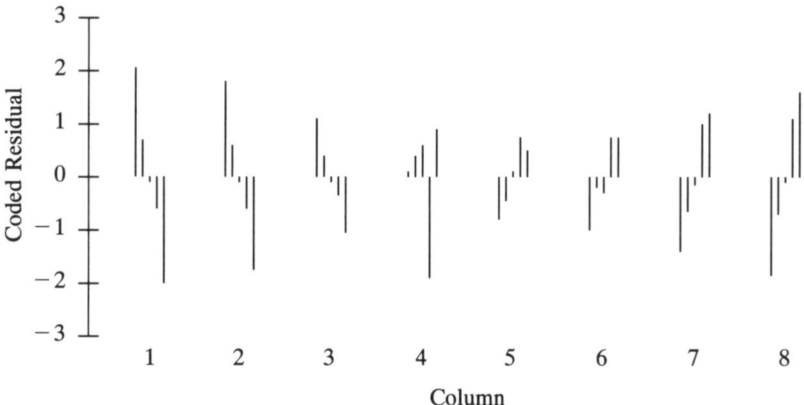

Figure 14.9. Residuals from additive fit. Data of Table 14.10.

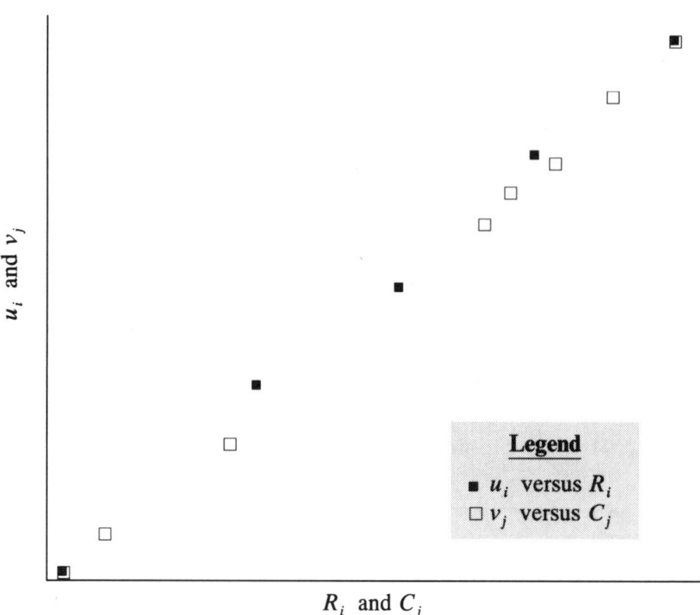

Figure 14.10. Diagnostic plot. Data of Table 14.10, with corrected cell (4, 4).

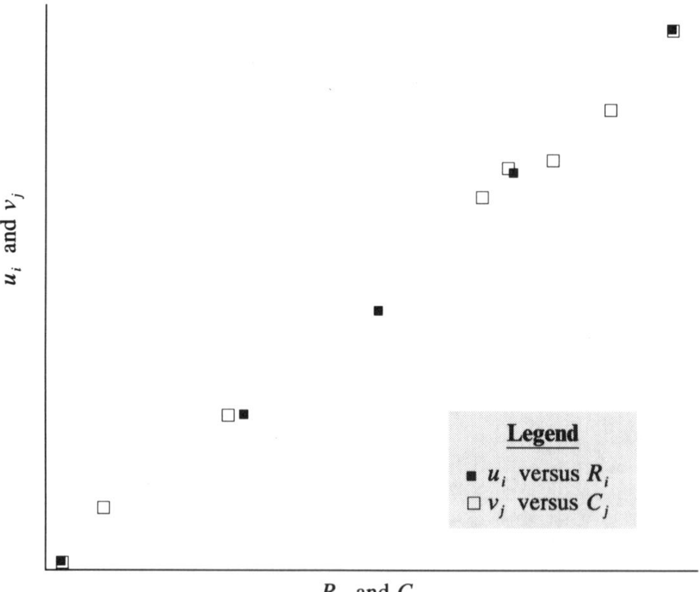

Figure 14.11. Diagnostic plot. Data of Table 14.11.

The residuals d_{ij} exhibited in Table 14.12 are seen to be very small, indicating a very satisfactory fit. The parameters R_i, C_j, u_i, and v_j are listed in Table 14.13, and the coefficients for a QFP fit to the quantities are presented in Table 14.14. Figure 14.12 shows the QFP fits, all of which are monotonic in the range of the experiment. Using these fits, we obtain the residuals shown in Table 14.15. Comparing Tables 14.12 and 14.15, we see that the QFP fitting has increased the residuals by a factor of approximately 10. We have succeeded in obtaining an empirical representation of the data within the ranges of pressure and wavelength of the experiment. We conclude that the data in Table 14.11 are well behaved and that we have been able to represent them by a simple empirical equation [$Eq.$ 14.4] which lends itself to interpolation everywhere in the ranges of pressure and wayelength of the experiment.

Table 14.12. Residuals from Additive-Multiplicative Fit [Eq. (14.4)] (Data of Table 14.11)

0.0001	0.0001	−0.0001	−0.0004	0.0005	−0.0003	−0.0004	0.0005
−0.0002	0.0003	−0.0004	0.0010	−0.0013	0.0005	0.0007	−0.0006
0.0004	−0.0006	0.0002	−0.0003	0.0009	−0.0006	0.0000	0.0000
−0.0007	−0.0002	0.0012	−0.0003	−0.0002	0.0008	0.0002	−0.0004
0.0004	0.0003	0.0008	0.0001	0.0001	−0.0004	−0.0001	0.0004

Table 14.13. Parameters for Additive-Multiplicative Fit [Eq. (14.4)] (Data of Table 14.11)

Rows			Columns		
Label	R_i	u_i	Label	C_j	v_j
1	1.522493	−0.631675	6678.149	3.059795	−0.559472
			6438.470	3.141196	−0.458833
255.4	2.816494	−0.283290	5875.618	3.385193	−0.275024
			5085.820	3.889396	0.130067
480.9	3.767619	−0.040907	5015.675	3.940394	0.191524
			4921.929	4.020994	0.204986
756.1	4.731119	0.301570	4799.920	4.133792	0.308348
			4678.160	4.254394	0.458397
1119.9	5.802997	0.654306			
$M = 3.728144$					

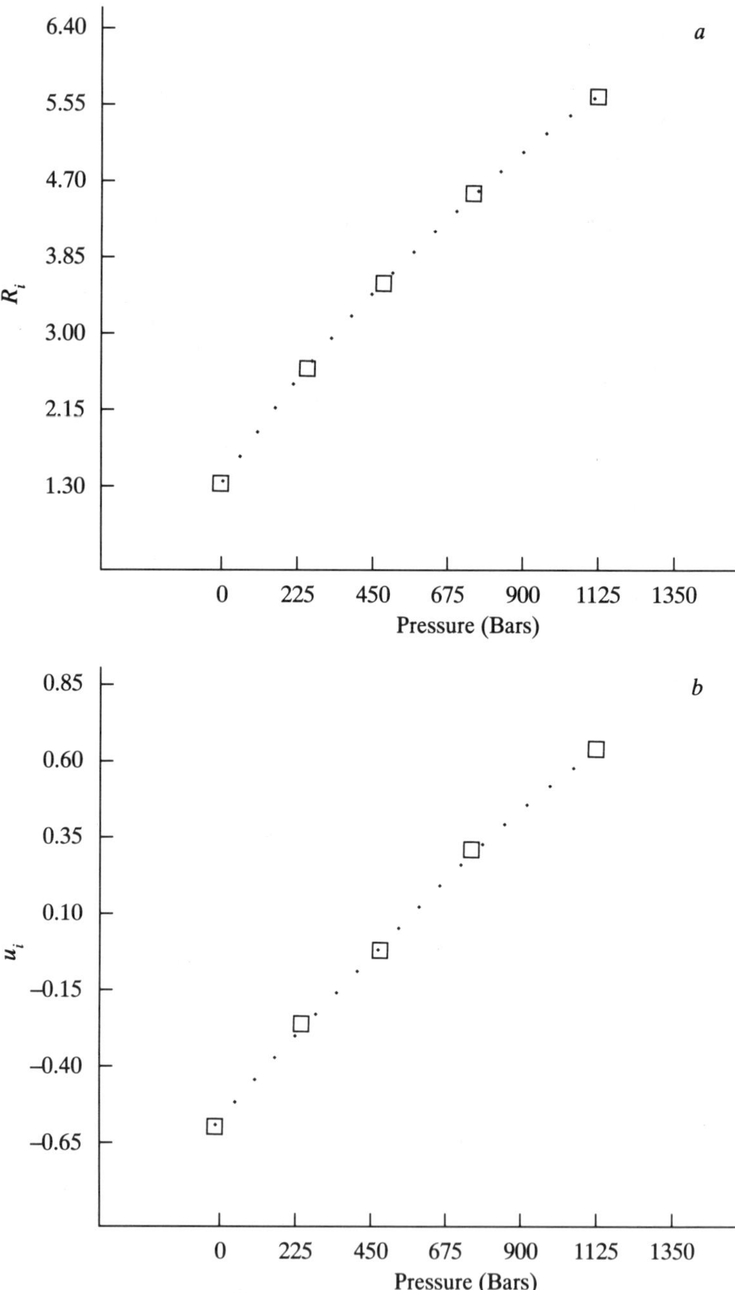

Figure 14.12. QFP fits for R_i, u_i, C_j, v_j. Data of Table 14.11.

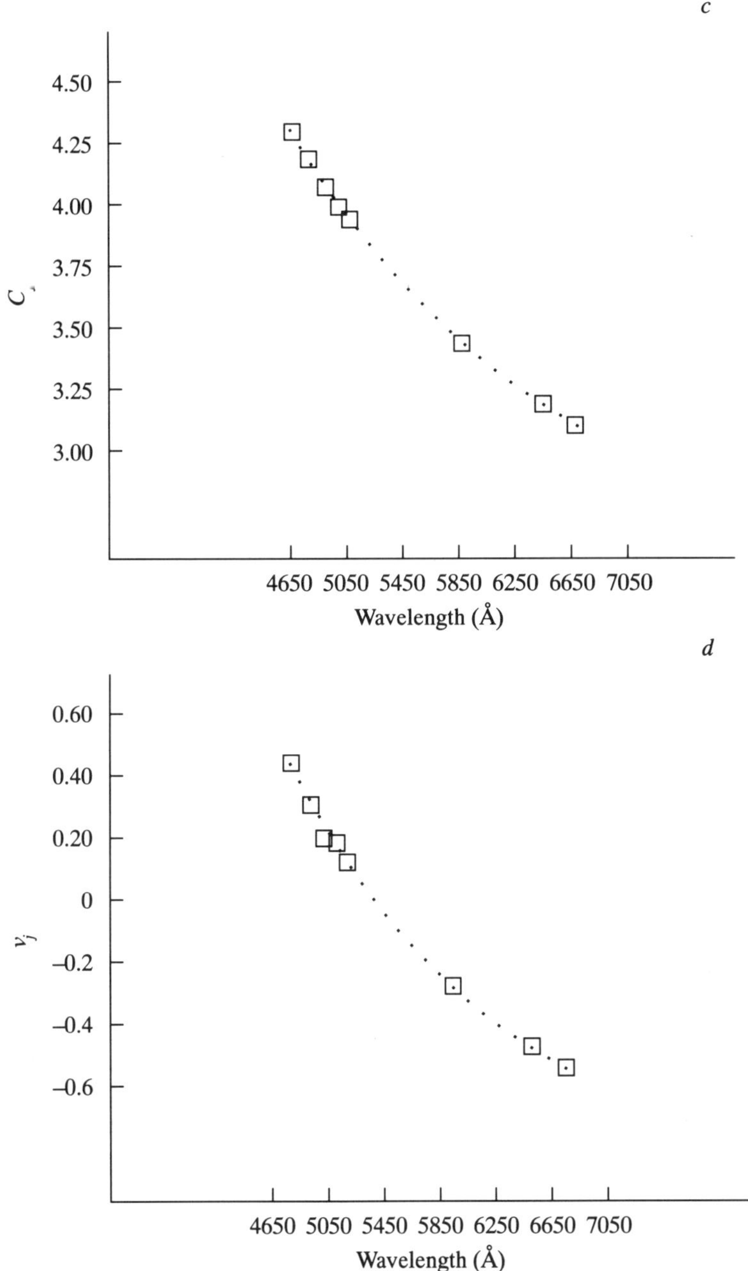

Figure 14.12. *cont.* QFP fits for R_i, u_i, C_j, v_j. Data of Table 14.11.

Table 14.14. QFP Parameters for Parameters of Table 14.13

QFP Parameter	R_i	u_i	C_j	v_j
$\bar{x}$	522.66	522.66	5436.719	5436.719
s_x	869.7648	869.7648	2069.692	2069.692
α	-0.2	1.2	$-2.$	-1.2
A	-0.879083	-1.101097	5.961484	2.689818
B	7.3362777	1.199799	-1.744596	-6.836489
C	0.126094	-0.143802	-11.89620	2.023650
$\theta = 0.101142$				

Table 14.15. Residuals from QFP Fitting of All Parameters (Data of Table 14.11)

0.0018	-0.0022	-0.0002	0.0025	-0.0035	-0.0003	0.0004	-0.0006
-0.0010	-0.0032	-0.0017	0.0035	-0.0051	-0.0011	0.0005	-0.0010
0.0065	0.0033	0.0054	0.0072	0.0023	0.0015	0.0034	0.0042
-0.0057	-0.0060	-0.0032	-0.0011	-0.0065	-0.0064	-0.0054	-0.0029
0.0015	0.0017	0.0013	0.0055	0.0004	-0.0030	-0.0004	0.0043

14.5 Potassium and Chlorine in Serum

We discuss again two interlaboratory studies in order to contrast them. Table 14.16 shows the data for potassium in serum (National Bureau of Standards, 1979). Six laboratories participated, each making duplicate measurements on each material on each of 2 days.

For the moment we can consider the experiment as consisting of 12 rows and 5 columns, with duplicates in each cell. Figure 14.13 is an h plot and Fig. 14.14 is a k plot. Neither shows any striking features. As expected, the laboratories show systematic differences from each other. There appears to be some similarity within each laboratory between the 2 days, with the exception of laboratory 1, for which day 2 shows strikingly low results. Table 14.17, which shows the analysis of variance of residuals, clearly indicates the superiority of the row-linear model over the others. Figure 14.15 is a plot of the standard deviations of repeatability and reproducibility as functions of level, ignoring the presence of days and considering the 12 rows as 12 laboratories.

Table 14.18 shows the results of the inter laboratory study of the determination of chlorine in serum. The design of the study is similar to that for potassium, except that 10 laboratories participated. Figures 14.16 and 14.17 represent h and k values, respectively. Table 14.19 shows the analysis of variance of residuals. It points definitely toward a row-linear structure. The h

Table 14.16. Potassium in Serum (K, mmol/liter)

Lab	Day	Material				
		A	B	C	D	E
1	1	1.349	2.629	4.428	5.574	7.350
		1.359	2.626	4.420	5.588	7.395
	2	1.261	2.456	4.285	5.346	7.082
		1.288	2.475	4.205	5.349	7.112
2	1	1.375	2.582	4.388	5.573	7.423
		1.372	2.528	4.355	5.548	7.390
	2	1.398	2.586	4.378	5.573	7.425
		1.501	2.578	4.549	5.569	7.409
3	1	1.269	2.474	4.356	5.517	7.461
		1.235	2.467	4.289	5.505	7.474
	2	1.206	2.441	4.211	5.553	7.465
		1.205	2.388	4.204	5.544	7.463
4	1	1.294	2.514	4.224	5.460	7.434
		1.292	2.484	4.168	5.506	7.382
	2	1.240	2.496	4.312	5.518	7.294
		1.276	2.394	4.264	5.406	7.262
5	1	1.269	2.543	4.178	5.463	7.308
		1.292	2.588	4.280	5.424	7.415
	2	1.274	2.462	4.295	5.451	7.359
		1.250	2.463	4.277	5.494	7.366
6	1	1.377	2.503	4.271	5.467	7.424
		1.430	2.482	4.241	5.513	7.390
	2	1.291	2.577	4.320	5.497	7.350
		1.344	2.582	4.309	5.487	7.403

Table 14.17. Analysis of Variance of Residuals (Data of Table 14.16)

Structure	Sum of Squares of Residuals	Degrees of Freedom	Mean Square
Row-linear	0.06646	33	0.002014
Column-linear	0.15089	40	0.003772
Concurrent	0.15262	43	0.003549
Additive	0.15370	44	0.003493

Table 14.18. Chlorine in Serum (Cl, mmol/liter)

Lab	Day	Material				
		A	B	C	D	E
1	1	79.53	94.70	102.38	108.57	117.21
		80.05	94.77	102.65	108.47	117.74
	2	80.32	93.57	103.20	107.91	117.28
		80.14	95.45	102.31	108.16	117.64
2	1	77.00	91.60	99.60	106.62	116.10
		76.75	90.70	101.70	104.90	113.90
	2	76.36	92.10	99.95	104.52	114.20
		77.94	92.66	99.95	106.10	115.00
3	1	79.37	94.88	101.90	108.00	117.71
		79.95	94.28	101.82	107.64	117.80
	2	79.49	95.37	102.70	108.26	118.45
		79.67	94.73	101.88	107.65	118.44
4	1	78.76	94.47	101.96	107.01	117.49
		79.20	95.31	100.00	107.20	117.68
	2	79.45	95.40	101.77	107.59	118.12
		80.36	93.57	102.05	107.29	115.49
5	1	78.05	94.33	102.01	106.92	116.64
		79.10	94.96	101.31	106.68	117.17
	2	79.43	94.40	102.07	107.21	115.79
		78.60	94.00	101.30	107.26	116.16
6	1	80.20	94.83	101.13	106.53	114.53
		79.50	94.10	100.87	107.23	113.50
	2	80.43	94.17	100.97	106.30	116.73
		79.53	93.70	100.17	106.07	116.30
7	1	78.35	94.50	100.40	105.20	115.60
		78.53	94.03	99.80	106.90	114.00
	2	80.10	93.10	99.41	107.29	116.95
		77.80	93.63	102.50	106.14	116.99
8	1	80.00	94.67	102.70	107.30	117.20
		79.70	95.00	102.00	107.00	116.50
	2	79.80	95.00	102.00	106.33	117.20
		79.50	95.00	101.90	107.20	117.00
9	1	81.50	96.60	103.30	109.60	118.90
		80.20	96.50	106.60	109.90	117.30
	2	80.40	95.10	101.40	107.00	117.30
		79.30	94.30	102.70	107.80	116.50
10	1	79.46	94.59	101.93	107.47	116.92
		79.54	94.76	101.99	107.52	116.89
	2	79.51	94.81	102.01	107.40	116.99
		79.39	94.68	101.97	107.23	116.86

Table 14.19. Analysis of Variance of Residuals (Chlorine in Serum, Table 14.18)

Structure	Sum of Squares of Residuals	Degrees of Freedom	Mean Square
Row-linear	13.8219	57	0.2428
Column-linear	23.0834	72	0.3206
Concurrent	23.2108	75	0.3095
Additive	23.4004	76	0.3079

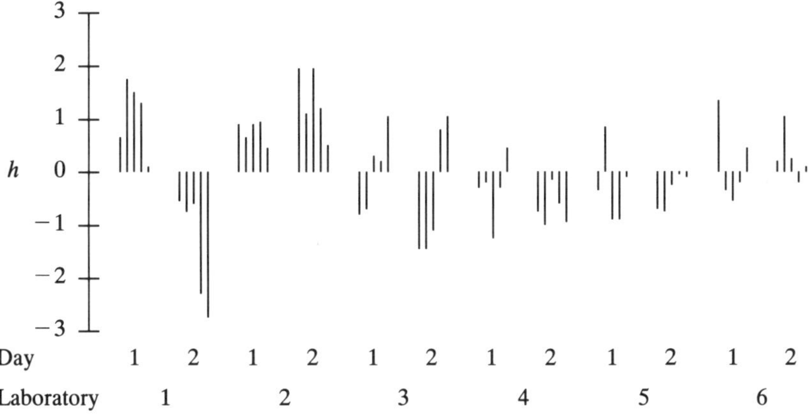

Figure 14.13. Graph of h-values. Data of Table 14.16.

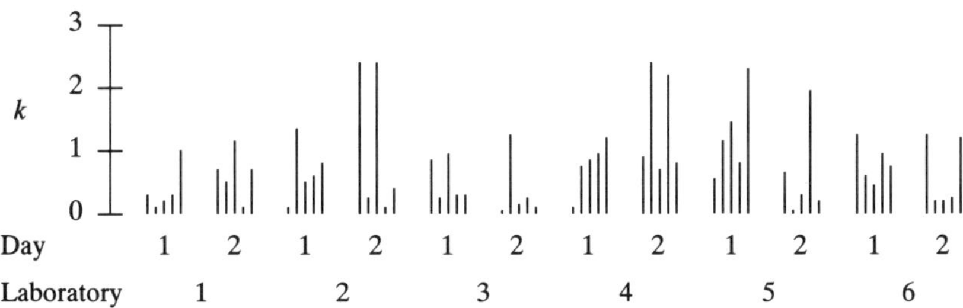

Figure 14.14. Graph of k-values. Data of Table 14.16.

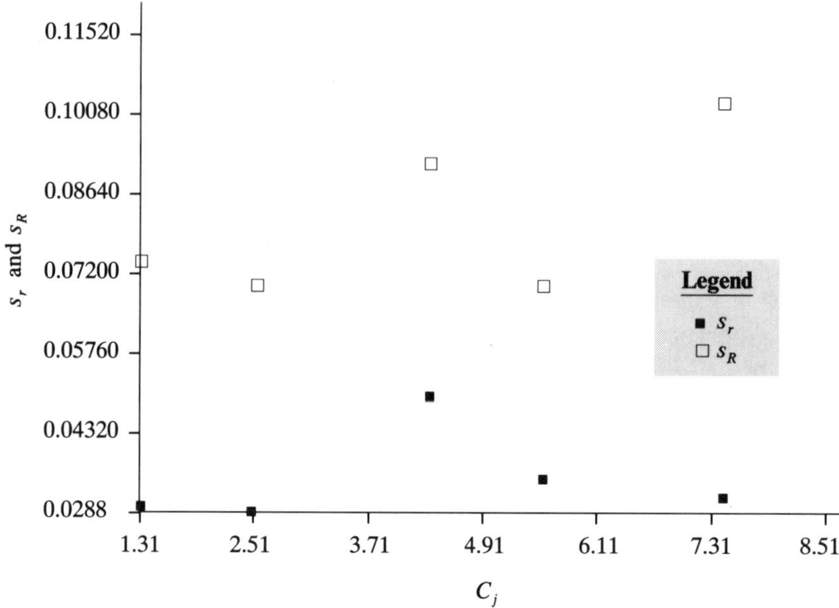

Figure 14.15. Graph of $S_r(j)$ and $S_R(j)$. Data of Table 14.16.

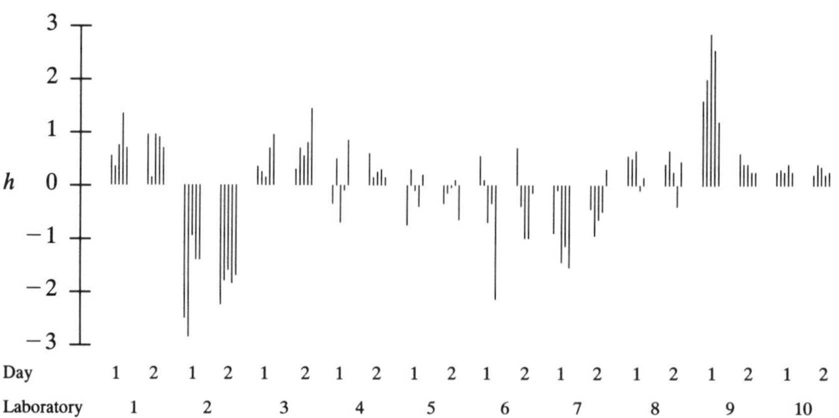

Figure 14.16. Graph of h-values. Data of Table 14.16.

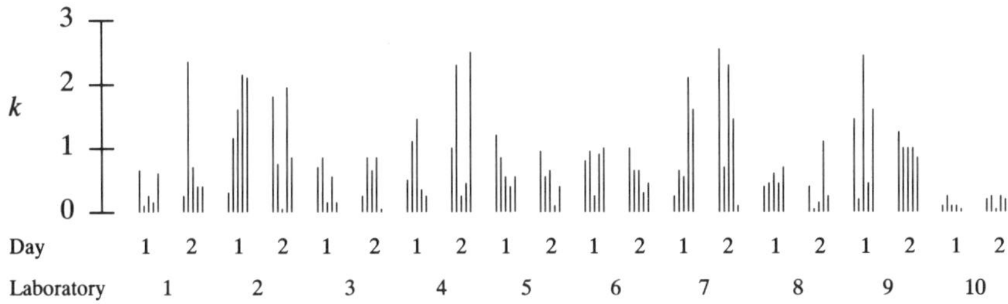

Figure 14.17. Graph of k-values. Data of Table 14.18.

plot, contrary to that for potassium, indicates fairly strong similarity between the results obtained on 2 days by the same laboratory. Again, however, there is an exception for laboratory 9. If we consider the 20 rows as 20 laboratories, we obtain the standard deviations exhibited in Fig. 14.18. The increase of standard deviation with level is slow, but it should be observed that the level increases only by a factor of about 1.5 from low to high.

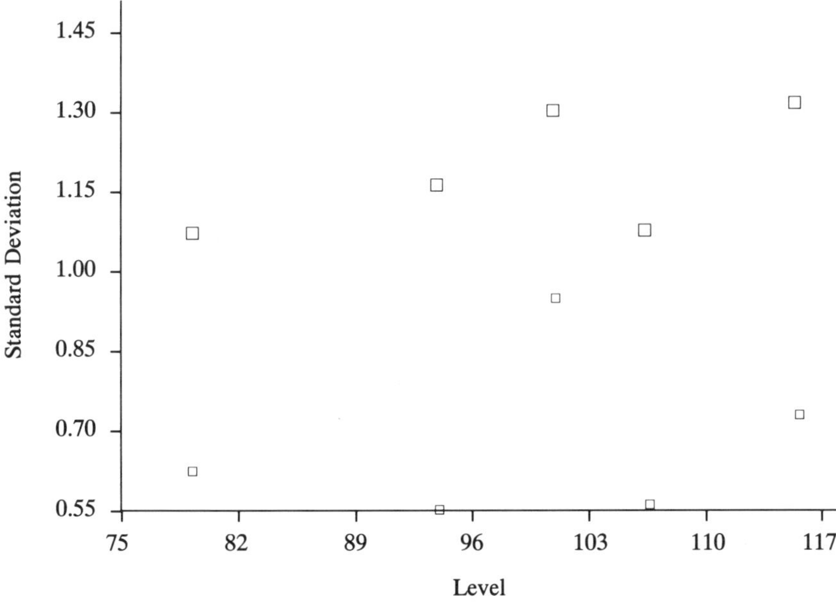

Figure 14.18. Graph of $s_r(j)$ and $s_R(j)$. Data of Table 14.18.

References

Bennett, C.A. and N.L. Franklin (1954). *Statistical Analysis in Chemistry and the Chemical Industry*. John Wiley & Sons, Inc., New York.

National Bureau of Standards (1979). *A Reference Method for the Determination of Potassium in Serum*. Special Technical Publication 260-63 p. 45, Table 9.

NBS (1979A). *A Reference Method for the Determination of Chloride in Serum*. Special Technical Publication 260-63, p. 38, Table 6.

Waxler, R.M., C.E. Weir, and H.W. Schampf, Jr. (1964). Effect of pressure and temperature upon the optical dispersion of benzene, tetrachloride and water. *J. Res. Nat. Bur. Stand.*, 68A, 489–498.

15

One-Way Tables
and the *h* Statistic

15.1 Introduction

The *h* statistic is not new. It has been used in a number of ways under the name *z*-score. By and large, its users have not fully realized to what extent a plot of *z*-scores can be a powerful tool of data analysis. Of course, before the widespread availability of computers, plotting was a tedious procedure.

Today, a computer can make a graph in a fraction of a second. We should take advantage of this circumstance for the purpose of analyzing data in complete detail. We will show, in this chapter, that this is possible not only for two-way tables as we have shown in the previous chapters but also for what is known as a "one-way" table.

15.2 One-Way Tables

A one-way table generally consist of measurements of a particular, single characteristic that fall into groups in such a way that in each group the measurement is repeated a number of times. An example is provided in Table 15.1 which deals with the vapor pressure of gold as measured by 10 laboratories (groups), each of which made several repeat measurements (laboratory 6 subrnitted no measurements). The number of replications within groups varies from group to group as seen in the table. It is customary to apply to such data a technique known as "one-way" or "within-between" analysis of variance. However, in such an analysis we have again a pooling of within-laboratory variances; this pooling is not always justified. The *h* plot is an excellent procedure for gaining an understanding of one-way tables.

In interlaboratory studies that can be represented by two-way tables we have used *h* in such a way that the standardization was carried out separately for each level. This was done because, generally, the levels in an interlaboratory

Table 15.1. Heat of Sublimation of Gold

Lab	Measurements						
1	89664	90003					
2	87176	91206	92638				
3	85924	84868	90142				
4	89133	85876	89815				
5	88508	87758					
6	–						
7	88464	85015	85687	89500	88911	86480	90212
8	86862	86300	87041	86927	86931		
9	87338	87702	91092	90977			
10	81981	98741	77563	84593			
11	98536	99568	70700	77335	89209		

Source: Mandel and Paule (1970).

study cover a wide range. That is not the case for the data in Table 15.1, all of which measure the same, single quantity (the Second Law Heat of Sublimation at 298°K of gold). Therefore, in this application we use only a single standardization to obtain the *h*-values. More specifically, let $\bar{x}$ be the average of all the data in the table, and let s be the standardization deviation of *all* these data (ignoring groups). Represent by y_{ij} the *j*th observation in group i. Then

$$h_{ij} = (y_{ij} - \bar{x})/s.$$

Thus, in this case *h* is simply a linear function, the same for all values of y_{ij}.

Figure 15.1 is a plot of the *h*-values by groups (laboratories). The groups do not all have the same number of repeat measurements, but this in no way affects the validity of the plot.

Figure 15.1, which is a complete representation of the data without any prior assumptions, shows at a glance the nature of the quantitative measures that can be obtained (see Section 15.3). Clearly, there is in our case no "within-laboratory" variability applicable to all laboratories. Laboratories 10 and 11 are appreciably more variable between replicates than the other laboratories. Table 15.2 shows, for each laboratory, its average and its standard deviation among replicates.

Because of the illegitimate pooling of within-laboratory variability, the customary one-way analysis of variance is not valid for this example. It is, however, possible to compute the necessary components of variance even in this case (Mandel and Paule, 1970). Details are found in Section 15.3 of this chapter.

Table 15.1 presents the so-called Second Law Heat of Sublimation at 298°K. Another method is the Third Law Heat of Sublimation at 298°K, for which the

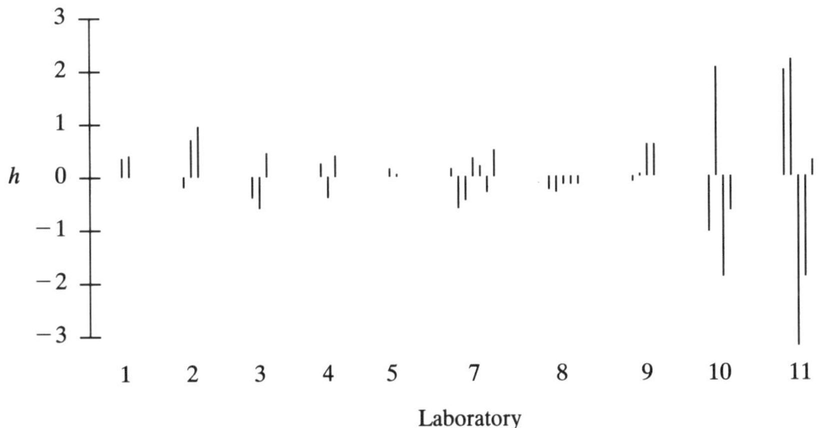

Figure 15.1. Graph of *h*-values. Data of Table 15.1.

data are given in Table 15.3. We can treat both sets as a single entity, standardizing in terms of the overall mean (both sets united) and standard deviation. The graph of *h* values is shown in Fig. 15.2. The graph is highly instructive: We see at once that the third-law data are far more reproducible within laboratories than the second-law data. We also see that laboratories 10 and 11 are much more variable than the others for the second-law data and that the same is true for laboratories 8, 9, and 10 for the third-law data. Further-

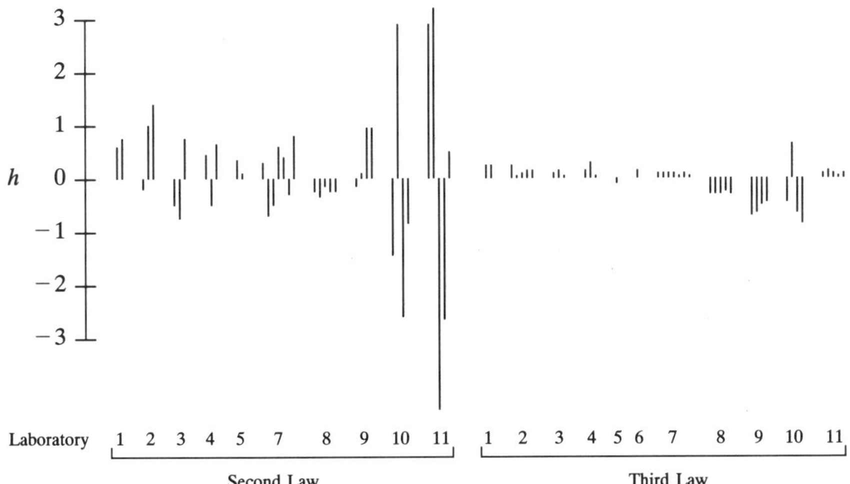

Figure 15.2. Heat of sublimation of gold (Tables 15.1 and 15.3).

Table 15.2. Heat of Sublimation of Gold

Lab	Average	Standard Deviation
1	89833.5	240
2	90340.0	2832
3	86978.0	2791
4	88274.7	2105
5	88133.0	530
6	–	–
7	87752.7	2014
8	86812.2	293
9	89277.2	2035
10	85719.5	9153
11	87069.6	12796

Table 15.3. Heat of Sublimation of Gold. Third Law

Lab	Measurements						
1	88316	88320					
2	88425	87626	87747	87975	88120		
3	87786	88108	87477				
4	88142	88566	87514				
5	87208						
6	88068						
7	87912	87917	87882	87791	87638	87845	87489
8	86653	86517	86570	86691	86531		
9	85047	85304	85679	86016			
10	85984	89821	85191	84624			
11	87911	88041	87911	87635	87693		

more, we see that there is apparently no systematic bias for any laboratory for the second-law data but that laboratories 8, 9, and 10 do show a negative bias in the third-law data.

Table 15.4 shows average result and standard deviations for each laboratory by both methods. To attempt to "weight" values by number of replicates would be totally inappropriate because "number of replicates" is only one of the variables affecting the results. Table 15.4 leads to no reliable conclusions regarding systematic differences in level between the second-law and third-law results.

Through application of the theory outlined in Section 15.3, we could obtain quantitative measures. It would have been preferable to find the causes for the greater variability or of the bias of these laboratories. Because this was not done, we will not perform any further analysis on these data.

Table 15.4. Heat of Sublimation of Gold. Summary of Results

Lab	Averages of Replicates		Standard Deviation of Replicates	
	Second Law	Third Law	Second Law	Third Law
1	89833.5	88318.0	240	3
2	90340.0	87978.6	2832	315
3	86978.0	87790.3	2791	315
4	88274.7	88074.0	2105	529
5	88133.0	87208.0	530	
6		88068.0	—	
7	87752.7	87782.0	2014	161
8	86812.2	86592.4	293	76
9	89277.2	85524.0	2035	407
10	85719.5	86405.0	9153	2345
11	87069.6	87838.2	12796	169

15.3 Iterative Weighting for One-Way Tables

Consider p groups of measurements y_{ij} where i identifies the group and j the replicate within the group.

We postulate the model

$$y_{ij} = \mu + L_i + \varepsilon_{ij}, \tag{15.1}$$

where μ is the true value of all y_{ij}, L_i is the bias of the ith group, and ε_{ij} is the error (departure from group population average) of measurement j in group i.

We define the group average for group i as

$$\bar{x}_i = (\Sigma_j y_{ij})/n_i , \tag{15.2}$$

where n_i is the number of replicates in group i.

Let σ^2 represent the variance of the ε_{ij} in group i. Then the variance of $\bar{x}_i$ is

$$\mathrm{Var}(\bar{x}_i) = \sigma_L^2 + (\sigma_i^2/n_i) . \tag{15.3}$$

The term σ_L^2 represents the variance of the L_i. The correct weight, w_i, for $\bar{x}_i$ is not $[1/(\sigma_i^2/n_i)]$ as in classical analysis of variance, but rather

$$w_i = [1/(\sigma_L^2 + (\sigma_i^2/n_i))] \tag{15.4}$$

Consequently, the best estimate for μ is the *weighted* average of the $\bar{x}_i$; representing it by $\hat{\mu}$, we have

$$\hat{\mu} = [(\Sigma_i W_i \cdot \bar{x}_i) \, \Sigma_i w_i]. \tag{15.5}$$

The mathematical problem is that σ_L^2 is unknown, so that the w_j are also unknown. Our solution to this problem is based on the fact that the expected value of the variance of a weighted average is unity, provided that the weights are the absolute (true) weights (and not a set of values proportional to them). Thus, the expected value of

$$\Sigma_i w_i \cdot (\bar{x}_i - \tilde{x})^2 \,]/(p-1), \tag{15.6}$$

where

$$\tilde{x} = \Sigma_i (w_i \cdot \bar{x}_i)/\Sigma_i w_i \tag{15.7}$$

is unity. We equate the sample estimate (15.6) to its expected value, unity:

$$\Sigma_i [\, w_i \cdot (\bar{x}_i - \tilde{x})^2 \,]/(p-1) = 1. \tag{15.8}$$

Denoting by F the difference

$$F = \Sigma_i [\, w_i \cdot (\bar{x}_i - \tilde{x})^2 \,] - (p-1), \tag{15.9}$$

we write, in accordance with Eq. (15.8),

$$F = 0. \tag{15.10}$$

The mathematical problem is to find σ_L^2 in Eq. (15.4) such that Eq. (15.10) is satisfied.

We solve it by iteration. We start with an initial value of σ_L^2, substitute it into Eq. (15.4), and calculate F [*Eq.* (15.9)]. If the value is not zero, we make a correction to it, using the Gauss–Newton procedure. This process is repeated until F is close enough to zero. This provides the best value for σ_L^2 and, consequently, for all w_i.

The iterative procedure is readily programmed on the computer.

Futher details can be found in the two cited references.

References

Mandel, J. and R.C. Paule (1970). Interlaboratory evaluation of a material with unequal numbers of replicates. *Anal. Chem.*, 42, 1194–1197.

Paule, R.C. and J. Mandel (1989). Consensus values, regressions, and weighting factors. J. Res. Nat. Inst. Stand. Technol., 94, 197–203.

16

How Useful Are Models?

16.1 Introduction

Much of this book is concerned with fitting a mathematical model to a set of experimental data. The statistical literature, at large, is even more concerned with what is called "modeling." As is usual with practices that are pursued with exaggerated zeal, modeling has been tacitly accepted as highly virtuous and, to this author's knowledge, its usefulness has never been questioned, nor have its limitations ever been stated.

At the risk of discussing the usefulness of a very popular doctrine, we will attempt to weigh the advantages and possible disadvantages associated with models.

The question is, What have we accomplished by fitting a mathematical model to the data resulting from an experiment?

16.2 Data Reduction

One obvious advantage is that a model can be a vast "data reduction" process. Thus, a table of say, 18 rows, 5 columns, and triplicates in each cell can sometimes be represented by a simple mathematical relation between a relatively small number of parameters. In this case, there are 270 data points, and they could possibly be reduced to 33 parameters or less.

Although this sounds like a great advantage, the question arises as to whether the 33 parameters really tell us as much as the 270 original data. In other words, can *all* the inferences that can be drawn from the original data also be drawn from the 33 parameters? The answer is negative. As a simple example, suppose that we fit a straight line to 20 pairs of x- and y-values. By means of the equation

$$\hat{y} = a + bx,$$

we have replaced the 40 values by a mathematical relation giving us an *estimate* of y for any value of x in the range of the table. But clearly, an $(x,$

$\hat{y}$) pair does not give us as much information as a corresponding (x, y) pair. Thus, although we can draw an uncertainty ellipse (see Chapter 10) about the (x, y) points, we cannot do this with the $(x, \hat{y})$ pairs. Thus, "fitted" values do not entirely replace the original values in usefulness. Against this disadvantage, we have the advantage that the relation

$$\hat{y} = a + bx$$

allows us to estimate y for an infinity of x-values that are *not* included in the original data set.

16.3 Prediction

A small set of data with quantitative predictor variables, such as the data of Table 11.1, is an excellent illustration for cases where the model is truly useful. Here we had a total of 41 measurements. The model that we fitted allowed us to predict the specific volume for *any* value of temperature and *any* value of pressure in the range of the table for each of two rubbers.

To cite another type of example, it is very likely that a relatively small number of measurements of volume would allow us to construct a very large table of predicted values for numerous combinations of pressure and temperature in an equation of state situation.

16.4 Linear Structures

The many examples of data sets that have been fitted by row- or column-linear models in the previous chapters of this book, and for which the labels were qualitative, are more problematic.

On the one hand, a mathematical model is a nice summarization of the data and generally gives us insight into the structure. On the other hand, it would be presumptuous to declare that the model represents the *true* relationship between quantities involved in the data set.

Some inferences can be drawn from the model, but there are also inferences that require that we look at the data themselves.

As has been pointed out several times in this book, outliers are *not* necessarily trash that has to be discarded. They may sometimes be eliminated from the model, but they often tell us important facts about the work in the laboratories that produced the data.

16.5 Minimal Assumptions Analysis

It is not generally recognized in the statistical literature that much understanding can be gained from data without fitting any model. As examples, we mention our h, h', and k graphs which require no model. We also call attention to our treatment of interlaboratory data (see Chapter 10) which calculates for each level a repeatability and reproducibility standard deviation, and then plots these against the level. The procedure for doing this is one-way analysis of variance at each level, a robust technique that is not entirely free of, but requires only a minimum of, prior assumptions. One of these is the equality of within-laboratory (repeatability) variances for all laboratories—an assumption that can be verified (or, often, contradicted) by a k graph. Even if this assumption turns out to be untrue (as often it does), the calculation of the standard deviations is generally not seriously impaired. Analyses of this type should always be accompanied by h and k graphs, so that any reader can evaluate what has been done. It should be noted that our one-way analysis of variance in interlaboratory testing does *not* include tests of significance, which would require assumptions about the statistical distribution functions of the data and which are unnecessary for our purpose.

16.6 Analysis of Variance

Analysis of variance (Anova) is considered by many as the ultimate "exact" method for analyzing data. It is nothing of the sort.

Anova is based on the mathematical fact that a sum of squares of numbers can be partitioned into several sums of squares, the sum of which is equal to the original sum of the squares. This is a mathematical identity, true for *any* set of numbers arranged in some systematic way. For example, if you fill Table 16.1 with any set of 12 real numbers, denoted by y_{ij} for the entry in the ith row and jth column, then the mathematical identity shown in the table is valid. This is totally obvious because upon algebraic simplification, the equation becomes $\Sigma_i \Sigma_j y_{ij} = \Sigma_i \Sigma_j y_{ij}$. Then denote the four terms on the right-hand side of the equation in Table 16.1 respectively, by "correction term," "row sum of squares," "column sum of squares," and "interaction sum of squares." Associate with them "Degrees of Freedom" equal to 1, $(3 - 1)$, $(4 - 1)$, and $(3 - 1)(4 - 1)$, and divide each sum of squares by the corresponding degrees of freedom and call the ratio a Mean Square.

For this entire operation to become more than a computational exercise, the mean squares must be associated in a meaningful way with physical realities (row effect, column effect, interaction). At this point, it becomes imperative to formulate a model. Thus, Anova without a model is merely a computational exercise.

Table 16.1. Schematic for the Anova of a 3 × 4 Table

		j					
		1	2	3	4		
i	1 2 3			y_{ij}			$y_{i.}$
				$y_{.j}$			$y_{..}$

Equation (Identity):

$$\sum_i \sum_j y_{ij}^2 \equiv (3 \times 4 y_{..}^2) + 4 \sum_i (y_{i.} - y_{..})^2 + 3 \sum_j (y_{.j} - y_{..})^2 + \sum_i \sum_j (y_{ij} + y_{..} - y_{i.} - y_{.j})^2$$

We have shown in this book that, for real data, the term "interaction" very often corresponds to *no* physical reality and, in fact, is the sum of several subterms which *do* represent physical realities. This points to the adoption of new models, against the simple "additive" model represented by the identity in the table.

Our conclusion is that Anova in the simple form presented in most textbooks, unaccompanied by a thorough examination of the structure of data, is mere computation and cannot be called data analysis.

An important further consideration is that even if the model has been properly identified and the Anova identity is changed accordingly, all we get is a number of "mean squares" that can be intercompared by F-tests under certain distributional assumptions. Many of the fine points about the data have been essentially submerged into these mean squares which, as seen in Table 16.1, are based on *sums* (or averages) of the data rather than on the data themselves. Thus, Anova is, at best, an "overall, average" evaluation, lacking many details that may be of interest to us.

16.7 Graphical Results

It is obvious to the reader of this book that we strongly advocate graphical procedures for data analysis. We did not present a systematic procedure for making graphs because we believe that it is an ad hoc procedure based on the problem at hand. Our h, h', and k graphs should prove very useful in many problems. We strongly advocate that they be made whenever possible. The so-called Box plots (Tukey, 1977) widely advocated in the statistical literature, will seldom tell anything beyond what an h graph will tell, and often much less.

The failure to make a graph can lead to serious errors, resulting, for example, in all sorts of significant interactions, all of which are due to a single aberrant value.

16.8 Models and Logic

A model is a structure, conceived by the analyst, that is imposed upon the data. The model is not in the data, it is imposed "on the outside" on the data. But in all cases, an attempt is made to "fit" the data as closely as possible with the model. In a way, a model is a "smoothing" device. For example, fitting a curve to a set of experimental (x, y) points is an attempt to represent only the "smooth" of the points, ignoring the "rough" (Tukey, 1977). It is, therefore, to be expected that part of the information inherent in the data may be lost in the model. Surprisingly, this is often ignored by practitioners who, after having fitted a model, draw all conclusions from the fitted data, thus forgetting that they have lost part of the information, even when the model fits the data very closely.

We now discuss an example of real data to suggest that in some cases, perhaps even the majority of cases, one can draw very useful conclusions without the benefit of a model. Our analysis is predominantly graphical, and it is hard to see how any model would have given additional useful information.

16.9 The Brightness of Shrink-wrapped Paper

A study was made of sheets of paper that were kept in a shrink-wrapped book to see whether the shrink-wrapping of the book had any effect on the brightness of the paper. The study involved four "volumes" (books) of paper sheets. For each volume, 100 sheets from a single source were randomized. Of these, 50 sheets were allocated to controls and 50 to shrink-wrapping. The sheets tested were the 6 near the front of the volume, 12 in the middle, and 6 in the back of the volume. Thus, a gap of 13 untested sheets exists between front and middle and a similar gap between middle and back.

On each sheet, six measurements of brightness were made, but we will be concerned here only with the averages of these six replicate measurements. These are shown in Table 16.2.

It is at once apparent that the shrink-wrapping caused a decrease in brightness of about 11 units. We are interested in the effect of position (the $6 + 12 + 6$ positions of the sheets in the volume) on brightness, both in the controls and the shrink-wrapped, and on the difference between control and shrink-wrapped. The possible differences between the four volumes are also of interest.

Table 16.2. Brightness of Paper

Position in volume	Sheet	Control				Shrink-wrapped			
		Vol. 1	Vol. 2	Vol. 3	Vol. 4	Vol. 1	Vol. 2	Vol. 3	Vol. 4
Front	1	74.98	68.65	72.22	74.85	63.20	63.24	64.27	62.83
	2	75.48	72.43	73.75	75.32	63.08	63.00	63.95	62.47
	3	75.50	74.18	74.40	75.23	63.23	63.23	63.47	62.77
	4	75.40	75.01	75.02	75.35	63.43	62.87	62.93	62.87
	5	74.75	74.78	74.97	75.42	63.95	63.30	63.05	62.73
	6	74.90	75.10	75.18	76.03	64.12	62.47	63.35	62.92
Middle	1	75.33	75.85	75.55	76.63	64.27	63.37	63.65	63.18
	2	75.32	75.65	76.45	76.73	64.47	63.72	63.88	63.42
	3	75.53	75.65	76.10	76.60	64.13	63.80	63.50	63.70
	4	75.25	75.93	76.33	76.73	64.23	63.77	63.93	63.33
	5	75.48	76.21	76.47	76.37	64.53	63.33	64.12	63.72
	6	75.27	76.07	76.13	76.22	64.42	63.72	63.92	63.50
	7	75.63	75.93	75.97	76.77	64.48	63.93	63.77	63.42
	8	75.53	75.70	76.45	77.03	64.80	62.68	63.88	63.08
	9	75.27	75.63	76.22	76.88	64.18	63.72	63.50	63.42
	10	75.70	75.70	76.55	76.80	64.45	63.15	63.80	62.83
	11	75.63	75.65	76.47	76.45	64.35	63.87	64.17	63.05
	12	75.38	75.78	76.68	76.70	64.28	63.73	64.25	63.27
Back	1	76.65	74.15	75.15	75.78	64.02	63.62	63.40	62.80
	2	74.28	74.58	74.63	75.43	63.52	62.60	63.50	62.65
	3	73.93	74.47	75.17	75.40	62.83	62.78	63.15	62.02
	4	73.88	74.08	74.32	74.93	62.87	62.63	62.87	61.87
	5	74.05	73.77	73.98	74.47	62.62	62.13	62.42	60.87
	6	72.75	72.92	73.90	73.49	62.08	60.77	61.95	61.12

Here again, simple graphs are extremely helpful. Figure 16.1 is a plot of brightness versus position for both the control and the shrink-wrapped sheets. Different symbols are used to characterize the four volumes. The two plots are in the same scale. We see that position has a definite effect: increasing from front to middle, then decreasing from middle to back. This is true for all four volumes. Is the intensity of this effect the same for both groups? Figure 16.2 is a plot similar to Fig. 16.1, but is made on the *differences*: control minus treated.

If the effect of position is the same for control and treated, then it should cancel out in the difference. Figure 16.2 shows that it does, indeed, cancel out to a large extent for all middle and back positions, but the cancellation is not complete. In the front positions, the disturbance caused by the outliers in the controls does not allow us to draw definite conclusions about the position effect in the differences between control and treated.

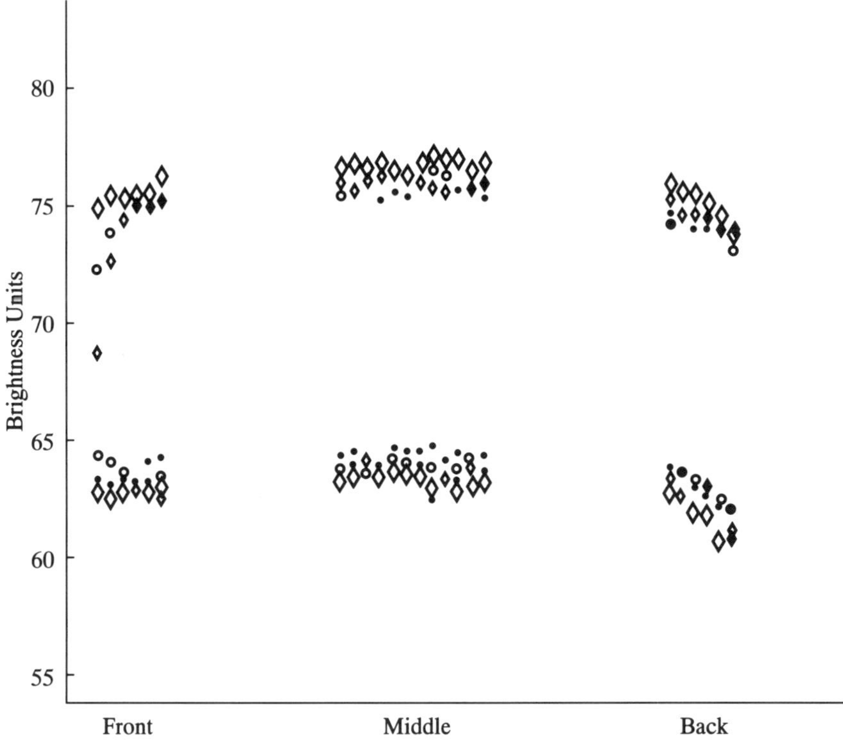

Figure 16.1. Brightness of paper. Six front sheets, 12 middle sheets, and 6 back sheets. Upper curve: controls; lower curve: shrink-wrapped. The four volumes are differentiated by symbols of increasing size from volume 1 to volume 4.

Figure 16.2 reveals further important features. Clearly, the difference control minus shrink-wrapped shows a definite volume effect: smallest for volume 1, somewhat larger for volumes 2 and 3, and largest for volume 4. This effect is most pronounced for the middle sheets and for the back sheets. For the front sheets, the effect is somewhat obscured by outliers in the controls (see Fig. 16.1). A similar structure holds for the effect of position on the difference between control and shrink-wrapped: For each volume, this difference seems to be unaffected by position within the middle and back positions; but in the front positions, the effect of outliers is evident and it disturbs the pattern. On the basis of these exploratory graphs, we can now draw some conclusions.

1. The shrink-wrapped sheets are about 11 units lower than the corresponding sheets.

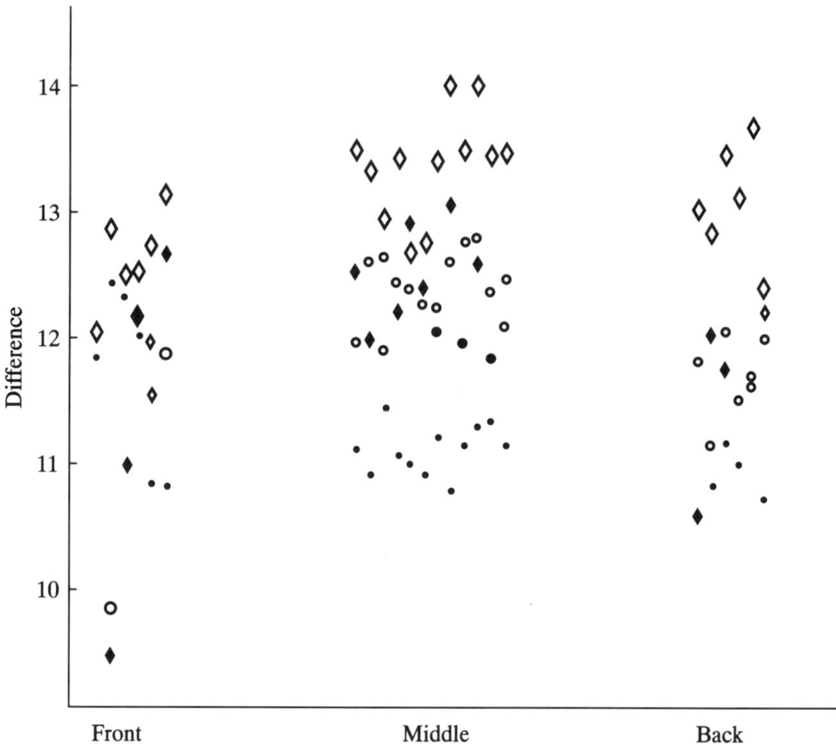

Figure 16.2. Brightness of paper. Differences between control and shrink-wrapped. Symbols as in Fig. 16.1.

2. Both in controls and the shrink-wrapped, there is a systematic difference between the four volumes.
3. Position has a definite effect on brightness, both in the controls and in the shrink-wrapped: The brightness increases from front to middle, and decreases from middle to back. This is true for all four volumes.
4. The controls show some outliers in the front sheets. This causes a disturbance in the differences between control and treated. These differences are, however, quite constant within the middle and back sheets, but they differ systematically from volume to volume. The differences are slightly lower in the back when compared to the middle positions.

Table 16.3 is a summary of the data. Of special importance is a comparison of the middle and back data as the front data contain outliers. First, we observe that for all four volumes, the middle averages are larger than the back averages. Using the standard deviations listed in the table and the fact that the middle

Table 16.3. Brightness of Paper. (a) Differences "Control Minus Shrink-wrapped". (b) Standard Deviations Among Replicates for Differences (a)

Position	Volume			
	1	2	3	4
(a)				
Front	11.6667	10.3400	10.7533	12.6017
Middle	11.0608	12.2467	12.4167	13.3325
Back	10.9250	11.5733	11.6383	13.0250
(b)				
Front	0.7133	2.6566	1.6190	0.3745
Middle	0.1957	0.4103	0.2447	0.4118
Back	0.3053	0.5692	0.3425	0.4417

averages are based on 12 values and the back averages on 6 values, we can calculate t-values according to the formula

$$t = \frac{\bar{x}_m - \bar{x}_b}{\sqrt{s_m^2/n_m + s_b^2/n_b}},$$

where m denotes middle, b denotes back, and n is the sample size.

The four t-values are 0.9924, 2.5819, 4.9689, and 1.4237. The evidence resulting from these values is not unambiguous: We are not certain that the differences obtained from the middle sheets will always be systematically higher than differences obtained from back sheets. At any rate, the difference between the two is small and of no practical significance. It would take many tables to convey the information provided by Figs. 16.1 and 16.2. Moreover tables with 24 rows (to represent the 24 positions) are almost impossible to interpret without a graph.

16.10 Conclusions

We have shown that graphs, when properly conceived, provide information that is both detailed and comprehensive about data. They have the further advantage of allowing the data analyst to decide what further analysis, if any, should be carried out. It is our contention that the use of computers in data analysis should mainly consist of making graphs appropriate to the data.

It is proper, after an analysis of experimental data, to ask whether the experiment has yielded useful information and whether it has raised questions that might be further investigated.

The brightness of paper data have yielded useful and precise information. Some of its results, such as the effect of position, are probably unanticipated. The questions raised by these data concern mainly the front sheets: Are they, in essence, more variable than sheets in the middle and in the back? Or, are the outliers experimental blunders? The answer to this question may not be of crucial importance, but it could have satisfied our desire to know more about the effect of shrink-wrapping.

Reference

Tukey, J.W. (1977). *Exploratory Data Analysis*. Addison-Wesley Publishing Co., Reading, MA.

Index